Richard Johnson

Protégé of Caroline Dormon

by

Loice Kendrick-Lacy

Copyright 2016

by

Loice Kendrick-Lacy

ISBN: 0-692-81321-7

All rights reserved

DEDICATION

This book is dedicated to the past, present and future generations of the Richard Lila Johnson and the Jessie Wade Fair Johnson families.

ACKNOWLEDGEMENTS

I wish to express thanks to Richard, Jessie, Rick and Ann for allowing me to have extensive interviews with them while gathering information for this book. A special thanks go to the following members of the Louisiana Native Plant Society for sending me information to use: Charles Allen, Gladden "Bud" Willis, Gail Barton and Linda Adrion. I am indebted to Jessie for giving me copies of several letters to include that were presented to Richard and her on the celebration of their fiftieth wedding anniversary.

Also, I appreciate James Durham for making available to me copies of the two speeches he made at the retirement party for Richard and Jessie held August 3, 2013. Both speeches have been included in this book. James, as well as other board members of the Foundation for the Preservation of the Caroline Dormon Nature Preserve, have furnished quotes to use. The other board members to be recognized are Gladden Willis, Julie Callihan and Chris Evans. I am grateful to Marion Bienvenu for permittting me to use her poem that she wrote in tribute to Sudie Lawton.

I wish to thank Amelia Hall, a student who visited Briarwood with Chris Evans and her fourth grade class from the NSU Elementary Lab School. Amelia wrote a glowing report of her experiences at Briarwood which has been published in this book. Lastly, I appreciate Mark and Amanda Stemmans for preparing the manuscript and photographs to be printed.

Table of Contents

Dedication iii
Acknowledgements v
Foreward ix
Richard's Birth and Early Years 1
Under the Wing of Caroline Dormon 5
Fishing and Hunting 9
Swimming, Church and Other Miscellany 13
Gypsies and Hoboes 19
Of Roses, Bullfrogs and Dipping Vats 21
Games, Other Activities and Pets 25
School Days and Graduation 29
Chores, Square Dancing and Parties 33
Richard Meets Jessie 37
Jessie's Birth, Family and Early Years 39
Engagement and Wedding of Richard and Jessie 41
Employment for Richard 45
Birth and Early Years of Rick and Ann 49
A Surprise Fiftieth Wedding Anniversary Celebration 59
Origination of Briarwood 75
Cotton Picking, Indian Lore and Miscellany 79
Helpers at Briarwood 83
Richard Becomes Curator of Briarwood 85
Retirement of Sudie Lawton as President of the Foundation 95
Formation of the Louisiana Native Plant Society 97
Wildlife at Briarwood 101
Student Helpers and Other Workers 109
Retirement Party for Richard and Jessie 111
Scheduled Events at Briarwood 115
Accomplishments and Plans for the Future 117
Notes From Members of the Dormon Foundation and LNPS 119
Conclusion 123

FOREWORD

This is a foreword which I feel honored to have been asked to write by time-honored friends Richard Johnson of Briarwood and the author Loice Lacy. It was a book that needed to be written about a man who has given so much of his life to this unique preserve of native plants and the accompanying other natural resources in this wonderful place. I love this place but I could never say that I have loved it as much as Richard and his wife Jessie who have so freely given more than forty years of their lives to Briarwood. Briarwood certainly sprang from the mind and efforts of Miss Caroline Dormon but it could never have been passed down to the rest of us in such a carefully-preserved state as it is now without the efforts of this wonderful couple. There is no better place to learn about our native plants and the way we and our wildlife are served by them than here. Richard and Jessie have been remarkably generous in spending time with me and with others who visit Briarwood. Someone once asked me when I thought Richard would retire and my response was "when he cannot open his front door." I hope that is many years away and my wish is that God's grace might descend on this couple and Briarwood; my distinct feeling is that it already has. Thank you, Loice, for doing something that very much needed to have been done.

Gladden Willis
Chair
Foundation for the Preservation of the Caroline Dormon Nature Preserve, Inc.

RICHARD'S BIRTH AND EARLY YEARS

The first time Richard saw Jessie, he knew she was the girl he wanted to marry. This man, who was curator of the Foundation for the Preservation of the Caroline Dormon Nature Preserve for forty-one years, is one to attain the high goals he sets for himself whether they be romantic or work-related.

Richard Lila Johnson, born November 7, 1928, was the sixth of the nine children of Rise Louis and Ellie Joanne Regions Johnson. Richard's siblings in the order of their birth were Julia Harvelia, Isobel, Willie Mae (she died at the age of five from pneumonia), Nettie, Mozelle, Louisa Lena, Jerry Willard and Marion Leslie.

Seated from left: Richard, Louisa, Jerry, Leslie
Standing from left: Mozelle, Nettie, Isobel, Julia

Richard says that his parents assigned each of their children to take care of a younger sibling. He added, "Julia drew the shortest straw and had to look after me." He remembers one time when he had done something Julia considered wrong, he crawled under the porch to escape her wrath. She got a long fishing pole and flogged him with it while he was under the porch.

When he was about five years of age, Richard had his first taste of what school could be like when attending a school Christmas party. At that time older siblings could take younger ones so Richard went with an older sister. As his gift there, he received a small mechanical caterpillar toy. Naturally he began playing with the toy immediately with disastrous results. It became entangled in the long hair of a little girl sitting in the desk in front of him. Richard has not admitted to whether or not this was intentional or accidental, but the girl's hair had to be cut to remove the toy. And the caterpillar was no longer functional.

As a result of the foregoing escapade, the teacher made Richard, the girl's brother and two other boys (all must have been considered guilty) stand at the blackboard for an extended time with their noses in circles drawn with chalk. It was at this time that Richard decided "I'd had about all the education I needed." Apparently Richard developed scholionophobia at an early age. No, I didn't make that word up; it defines his dislike of school.

Richard and his friends would make "darts" using the thorns of honey locust stuck in corn cobs. He would only throw the darts at trees, but in the summer right before he was to enter first grade, one of his friends killed the teacher's rooster with a dart. This got Richard off on the wrong foot with his teacher as she believed he was the one who had killed her rooster.

In the classroom, boys would place a bobby pin in a crack in a desk in front and use a barefoot toe to play a "tune" on it. The teacher usually couldn't tell who was guilty so Richard says he was never punished for this misdeed.

For "show and tell" at school, Richard once took a paper bag containing three hibernating bats. He placed the bag on the teacher's desk which was quite near the heater, causing the bats to begin movement as they warmed. When the teacher opened the bag to view the contents, she began screaming. The students were sent outside while the janitor opened the windows to shoo the bats out of the room.

To get to the Readhimer school he attended, Richard walked about one and a half miles down the railroad tracks or a mile through the woods, his preferred route. In the spring, he would pick native azalea blooms to take to his teacher. Some of his friends laughed at him for this practice but Richard says, "It helped my grades." Many of us realize that an apple for the teacher can be advantageous.

native azalea blossoms

Richard says he had only one teacher who really understood him. She encouraged projects that interested him, as he says to "keep him out of trouble." When he was about thirteen he and three of his classmates took a large table on which they put three inches of sand as a base for the construction of a facsimile of the Readhimer community. It depicted houses, a school, an auditorium, Baptist and Methodist churches, pond, fields, rail fences, pig pen, trees, shrubs, and contoured hills. And even an outhouse! They used green crepe paper for the trees, shrubs and grasses. Richard says the project was a real hit.

UNDER THE WING OF CAROLINE DORMON

Richard grew up under the wing of Caroline Dormon, working with her even as a child. He describes her as having been a "hands on" person and himself as her "gofer." Although he says she would sometimes scream at him, he says he still misses even that, pointing out, "She was my mentor. Miss Carrie helped raise me, more or less. I was a squirrel she couldn't get rid of."

On a tour with Richard during the annual picnic at Briarwood in April 2015, Cindy Walker says he told the group that many women, girls, and even some boys were afraid of Miss Carrie because she was "different," but that he never was. Cindy says he also told them about running up the trunk of a fallen tree that had lodged against another just to see how far up he could go.

Richard was only in the first grade when he began helping Caroline to gather smilax, mistletoe, holly and such to sell at a nursery in Shreveport. He says, "She took an interest in me at a very early age. She hired me to climb trees and pull mistletoe out of the tops. Back then, I could climb trees really well." He remembers that Miss Carrie would often yell at the robins for feeding on the berries of Briarwood's beautiful hollies that were one of her prized cash crops. When working among the plants, Caroline would frequently tell Richard, "Get down on your prayer bones, boy!"

Richard says that Caroline had what she called the gift of the wild things. "She had some sort of agreement with nature." According to Richard, Caroline's early interest in nature was due to the influence of her quail-hunting father. He says, "It was back in the times when a Southern gentleman went hunting on a mule and shot from it. He could see the bird dog pointing and he would ride out and shoot the quail. Caroline would go and pick up the bird to bring it back, and each time she'd bring back a wildflower. Her father would explain to her what the flower was and why it came to be here at Briarwood."

At the age of twelve Richard began working at Wild Gardens Nursery at Choctaw Bottoms helping dig the plants and bag them to be taken to Shreveport and Baton Rouge for sale. Part of his job was to get green sphagnum moss to pack plants like violets for shipping. He knew where to find the best holly to collect as well as mistletoe which was sold for Christmas decorations -- or to stand under all puckered up for a kiss. Caroline's sister Virginia was the driver of the vehicle used to deliver the plants. Richard says, "Miss Virginia was a fast driver; she didn't let the grass grow under her feet."

Richard had learned how to graft at the early age of fourteen. One of his specialties was grafting Japanese persimmons onto native rootstock. Once when a group from Louisiana State University had come to Briarwood, Caroline told them, "I want you to watch this boy. His grafts never fail." Well, this time they did, but perhaps it was because they were done in a drizzling rain. Fortunately, those from LSU never knew about the failure so Richard's statewide reputation remained unblemished.

As a child, Richard was allowed to go to Briarwood when there were no visitors there. When he was about seven he came up the trail and saw a long stretch limousine parked at Briarwood. Since this was bad timing for a visit, he ran on to the store out at the highway then breathlessly back home to tell his dad, "I saw a car as long as a school bus and the Negro driver was sitting up front outside the car!"

The driver chauffeured Josephine Henry and her mother, both from Pennsylvania, around the country collecting plants. One half of the vehicle had been converted to a weed wagon to haul the plants and presses. Since this was the 1930s, the driver slept in the limo, not in a hotel as did his two passengers. Richard says, "It knocked the props out from under me when I learned that the Negro held a Ph.D. from Pennsylvania State University."

One of the reasons Richard loved to visit Caroline and Virginia was so he could read their collection of National Geographic magazines and Raymond L. Dittmar's book on snakes. He and other family members being avid readers, he was allowed to "check out" the Dormon books but was asked to give a book report when returning them. And Miss Virginia's delicious ginger cake cookies were an added enticement for visits!

FISHING AND HUNTING

Richard never knew his paternal grandfather as he died when Richard's father was only twelve. Being the oldest of the children, his dad had to quit school to take a job as a water boy for the railroad to help support the family.

Richard describes his maternal grandfather as being "short, pigeon-toed and walked fast like a sparrow." His grandfather had told him that when he got big enough to carry his own pole, he'd take him fishing with him. That time came when Richard was six or seven years of age. Although the pine pole was really heavy for a small lad, it was great for catching buffalo in Saline Creek. The buffalo, or native carp, often weighed as much as eight or nine pounds and could be found in the curves of the creek where the water was sometimes fifteen feet deep. Richard says that this fish, named for the hump on its back, is delicious to eat.

The fishing tackle was carried in a syrup bucket which was later used for making coffee. No worry even if the bait was worms; the heat sterilized the bucket. However, bait was often simply unleavened dough balls. Poles were usually pine as bamboo was unusual in that area.

Once while he was just a youngster, Richard got a ladder and climbed up on the house to get a bunch of the lead heads from off the nails that were securing the tin roof as he wanted them for sinkers on his fishing line. He was so pleased with himself for having thought of this that he showed them to his mother. Needless to say, she was shocked as she knew removing the lead from the nails would cause the roof to leak.

Richard was flattered that he was the only boy that his grandfather would take fishing. Once on a trip with his grandfather and Uncle Jim, the men placed Richard between two cypress stumps. He stuck his pole in the mud and was sitting half asleep when something gave a big jerk on the pole. He locked his legs around one of the cypress stumps and yelled for help. His grandfather exclaimed, "I don't know what it is, Jim, but it's a bigun'!"

Richard landed a huge snapping turtle which was too heavy to carry. After cutting off the turtle's head, holes were then drilled in the shell with a pocket knife to insert a line to be able to pull it to his grandfather's house. Richard's grandmother and his Aunt Pearl dressed and cooked the turtle. He kept the shell to show off until it completely disintegrated. In telling this story to me during an interview, Richard said that he had recently seen on TV that a man who was bitten by a snapping turtle had to go to the ER for treatment.

At the age of about ten, Richard decided to do a little fishing in a fish pond that his older sisters, Julia and Isobel, had landscaped with plants around it. While trying for the fish, Richard backed into a prickly pear with disastrous results. At the time, he had only two pairs of overalls for school and the one he was wearing this day had to be discarded because not all the spines from the cactus could be removed.

With only one pair of overalls, Richard had to wear a dress belonging to one of his sisters each time his overalls were being washed. Richard's Uncle Sid came by one day and caught him wearing the dress; naturally his Uncle Sid never let him live it down. Jessie says this was poetic justice for trying to catch his sisters' fish.

Richard despised wearing overalls, saying they were "just like wearing a cotton sack." He says that when he was active, the galluses were always slipping off his shoulders. He was about in the fifth grade before his parents gave up and let him start wearing khakis. Jessie says that she hated wearing overalls as much as Richard and both of them were so glad to leave those overalls behind and wear something else. Richard was even more pleased when denim blue jeans became available. He says, "Now I put on something else to go to church."

As most country boys do, Richard often brought home wild game for his mother to cook. Usually she obliged him, balking only at cooking possums, but fortunately he had an Aunt Audrey who would. Richard said that even during the Great Depression, they always had plenty to eat whether it be from hunting, fishing or things that they grew. He recalled storing sweet potatoes in a huge pit for later use.

When Richard and his cousins would go possum hunting, they used torches made from pine splinters as a means of light in the darkness of night. Focusing the beam up into a tree top would reveal the shining eyes of any possum that had sought refuge there. In current times one would simply shine a light from an iPhone up into the tree, no need to harvest pine splinters.

SWIMMING, CHURCH AND OTHER MISCELLANY

Richard waxed nostalgic in the April 2013 issue of the Caroline Dormon Nature Preserve newsletter:

> As spring warms the good earth it always brings back memories of "the way it was." As we turned the earth with mules and plow the wonderful aroma of newly plowed soil greeted us. Walking barefoot through the field then was a sheer delight. We knew that in a few weeks the pond and creek waters would be warm enough for swimming. Of course all winter we would dare each other to take a plunge into the icy cold waters. Afterwards it was a race to reach the fire we had built on the bank. Those moments stolen away from field work were fleeting so we made the best of them. We children never had time to be bored.

Richard's dad couldn't swim but most of his children enjoyed it immensely. Richard said that he thought his dad was fearful of a drowning as that would mean one less for cotton picking. He said that families needed to be large to have extra hands in the field, adding, "Birth control and cotton picking machinery cut back on the number of children in a family."

Richard had an uncle and aunt who took all the kids to Saline Creek to swim each Sunday afternoon in warm weather. When their preacher learned about it he preached against it, putting a stop to what he called "mixed bathing." The following summer they had a new preacher named Roy Remont, a Cajun from Cutoff, Louisiana. Richard says, "Rev. Remont loved to swim so we all went!"

In addition to being the minister at Briarwood Baptist Church, Roy Remont also taught at Readhimer High School. Years later after he had left the area, he returned for a reunion. Richard took this opportunity to thank him for having approved of the swimming they all enjoyed so much.

There were leeches in the creek which were taken off using salt. This is not a recommended procedure now as the salt causes the leeches to regurgitate into the wounds. Remember Humphrey Bogart covered with leeches in the movie "African Queen?"

Roy Remont married a local Methodist girl, Betty Williams, who rode a mule named Tootsie everywhere she went. Betty's mother, a teacher at Readhimer, had once boarded with Caroline and Virginia before she married. Richard says Betty was a sweet girl with a great sense of humor, but that her brother Charles "could get into more mischief."

One day the school bus driver gave Charles a paddling and put him off the bus. The following morning Charles threw a live skunk into the bus, but Richard doesn't remember exactly how this worked out.

While an employee of LP& L, Charles was working on cutting sprouts of sassafras and persimmon on the right-of-way. He told those hauling off the sprouts to dump them in the yard of Woodrow Wafer as he wanted them to use as stakes. When Woodrow found out who had ordered the sprouts dumped in his yard, he was furious but didn't dare cross Charles, knowing his reputation. Charles was "getting even" with Woodrow for having picked on him off and on in the past.

As a child Richard remembers attending the Primitive Baptist Church of Readhimer with his maternal grandmother. She kept him close by her side so she could pull out a thimble she carried in her purse to thump him sharply on the head if he misbehaved. Another of his remembrances is that his Uncle Charlie would often go to sleep in church and snore loudly. At times the family went to other Baptist and Methodist churches. He says that as a youngster, he never thought the communion bread at any of them had enough salt in it.

At this time there was an annual ritual of washing of feet at many of the churches. A man would tie a towel around his waist to wash the feet of the men while a woman washed those of other women.

In the summer the benches of the Briarwood Baptist Church were moved out under a brush arbor for revivals; other churches also did the same. Coleman lanterns were used for lighting at evening services. Richard told me, "Miss Virginia (Dormon) drove with her legs crossed. One night she couldn't get them uncrossed in time and took out about three of the back benches under the arbor." The next meeting night the preacher announced to the gathered worshippers, "You'd better move on down to the front. Miss Virginia will be here shortly."

As children, Richard and about six of his cousins and some of his siblings would sometimes play having a church revival. They sang hymns by heart and Richard's older sister Nettie would baptize them in the sand. The converts would lie down in the sand and she would "immerse" them with it. One day Charlie Madden slipped up on them and began laughing at their activity, which greatly embarrassed Nettie. But Richard says, "She could make a Christian out of all of us when she became displeased."

Charlie Madden was called "Uncle Charlie" by the youngsters because he was in his forties when Richard was still a child. Everyone liked him for the humor he used to make friends. He sometimes played Santa, fitting the role well as he was jolly and had a big stomach, lacking only the beard.

On the way to the community store which was about three quarters of a mile from his house, Richard got to studying all the plants along the way. As a result of his early interest in the natural world, he taught his cousins and some of his friends about nature. He was even able to identify a tree for his dad that had been puzzling the older Johnson. Richard's dad was raised in the country near Sand Spur just west of Pollock, Louisiana and remembered having seen trees there that had extremely large leaves. Richard told him that they were big leaf magnolias (*Magnolia macrophylla*).

Jessie says that there are many big leaf magnolias at Briarwood, spreading from those for which Caroline had planted the seeds many years ago. She says that there is a problem with these trees because, as soon as they reach sapling size, male deer find them perfect for rubbing the velvet off their antlers. This leaves ugly scars where the deer rub down through the bark.

The writer Ada Jack Carter and her son David Snell, also a writer who was once editor of Life Magazine, made a visit to Briarwood when David was about four or five years of age. He recalls that he and his mother were riding with Virginia to Saline to get a block of ice. In relating this episode to me Richard said, "Miss Virginia could talk and possibly stay in the road." But this time she had an excuse as the steering wheel came off causing the car to leave the road to crash into a pine tree on Briarwood. The car's occupants were unhurt but the fender of the car was badly dented and water and steam were gushing from the radiator which had been caved in.

Terrified of what Caroline would say if she knew of the gaping wound on the trunk of the pine tree, Virginia had the others help her push the car back onto the lane and nudge it against another pine tree to make it appear the accident had occurred there. Then they used soil, branches and pine straw to conceal the wound on the damaged tree.

Richard described a major hurricane that came ashore at Morgan City in 1934 which killed seven persons. He said, "There was also a hurricane in 1935 or '36 that brought in a lot of exotic birds from the tropics, many of which were picked up dead. In this storm the hail was so big that it beat off all the upper parts of tree limbs. Dad was walking home from a house where he had picked up a quilt that the neighbor had borrowed previously. He used the quilt and a cotton basket to protect his head from the hail."

With the storm approaching, Richard's mother told Julia to put up the animals. The cows were put in the shed, the mules sought shelter on their own, but the horse just didn't want to be penned. You'd have thought he'd have exhibited more horse sense.

GYPSIES AND HOBOES

Tribes of Gypsies migrated in wagons to Louisiana in the fall each year, then in the spring going as far north as New York. In all there were around 10 wagons -- some only a one-horse wagon covered with lumber -- with about 30 adults and many children in the clan. The men and children walked, riding in the wagons only when it was raining.

The main winter location for the Gypsies was at St. Martinsville, but they also camped at the parish line of Natchitoches and Bienville under pine trees on their way to and from St. Martinville meeting. In their winter conference, leaders were elected. The king, who was the official in charge of the tribe, was a quite wealthy man. They worked at such odd jobs as repairing holes in pots and pans, sharpening knives, etc. Richard says the Gypsies had a reputation for being light-fingered, once prompting a local store owner to remark that he had to watch them or the women would carry off half the store under their full-skirted dresses. Gypsies were present in Louisiana from 1935 until the war stopped their migrations in 1941.

In the winter, hoboes built hogans along the tracks of the L&NW (the old Doodle from Natchitoches to McNeil, Arkansas). The hoboes traveled with their belongings tied to a stick they carried over their shoulders. Richard called the hoboes "nomadic survival experts" and described them as intelligent people but with a reputation for stealing chickens. He says many people were afraid of both the hoboes and the Gypsies but to him they were "free spirits like myself."

During the Great Depression there were many hoboes "riding the rails," a practice which became much less common with the beginning of World War II. To read about some of their astounding – and often poignant – escapades, google "hoboes in Louisiana."

OF ROSES, BULLFROGS AND DIPPING VATS

Richard says that his mother grew the best roses in the community. He remembers two big cabbage roses from which he loved to eat the sweetish-tasting blooms. With many kinds of roses, every day in the summer Richard had to draw two-gallon buckets of water out of the well to water each of them.

On wash day, which was once weekly in the winter and twice weekly in the summer, another job for Richard was drawing well water for the wash pot where the clothes were boiled and for two rinse tubs. The wash water was later used on the plants as even the soapy water was beneficial for the roses.

When I was interviewing them, Jessie said she simply couldn't wring out the clothes very well. Richard told her, "If you ladies had tried milking two cows every day you'd get your hands strong."

Both Richard's and Jessie's mothers swept their yards as did all the ladies of that era and each yard had at least one garden bed with Garrett Snuff bottles for a border. Very particular about the appearance of her yard, Richard's mother would say to her children as they would pass the homes of blacks, "If they can keep their places so spotless, you can do the same."

One reason for the practice of sweeping the yard was to help keep grass fires from reaching the house, while another was to prevent snakes from having a hiding place. Richard told me, "My dad had a fascination with snakes but my mother didn't go for that."

During his high school years, Richard helped as a volunteer at the community cafe, mostly on Saturday nights but at other times when needed. He would leave the keys in his coupe for other boys to use it to pick up their dates. When he discovered that some were putting casing head gas in it he told them, "Don't mess with my 'courtesy card' (meaning his siphoning hose). It's for emergencies only." He also found that gas was sometimes being stolen from him using the siphoning hose.

Roaten's Cafe

When Richard was a junior in high school, he and two other boys visited a friend who was staying home alone while his parents were away. The four boys killed and skinned about 25 bull frogs. Even after its head was cut off, one frog made two or three jumps across the kitchen floor. Richard showed the others how to touch a certain nerve to make a "dead" frog quiver. This repulsed the others so much that one boy said, "You may eat them, but not me." In the end, Richard was the only one who would eat the frogs.

Encountering seed ticks (freshly-hatched ticks in the larval stage) was to be expected if one wandered in the countryside. To get rid of them, Richard says they singed their legs by lighting a long piece of paper and moving it up and down them. This also would get rid of any chiggers present. An added result of the singeing was that the girls didn't have to shave their legs. During an interview, Jessie said that in later years a common practice was to grab a handful of sand and rub it up and down legs to rid them of ticks.

During this time the government had dipping vats where farm animals were required to be dipped to get rid of ticks that were causing tick fever. A theory has been presented that the plague that God sent against the Egyptians (see Exodus 9:1-7) was tick fever. Dipping in Louisiana began in the early 1900s. Southern livestock had resistance to tick fever but if not dipped before being shipped to northern states, they could carry the disease to the animals there that could seldom survive once infected.

Richard tells a story about once when a young friend of his, J.T. Taylor, had taken his animals to be dipped, he was accompanied by his herd dog. The ranger – an overseer at the vat – thought it would be funny to strike the dog with the electric probe used to urge the cattle through the vat. The dog jumped into the vat in its reaction to the hot stick. After retrieving his dog, J.T. grabbed the probe and hit the ranger over the head, knocking him down. Richard's great-uncle, Marion Regions, had this to say about the incident, "Two things: don't mess with another man's wife or his dog."

Richard says that there were many who objected to the government making dipping mandatory. He remembered that fifteen vats were destroyed one night in an act of vandalism. At another time two men near Pollock were killed because they were determined to continue dipping. Richard says that after being dipped, a cow's teats would be so sore from the kerosene used in the vat that she would respond by vigorously kicking which made milking a difficult task.

During this era the government furnished large bottles of cod liver oil to families as it was considered nutritious and thought to prevent rickets and various other diseases. Richard's younger brother Jerry, who was born deaf presumably from his mother having had German measles while pregnant with him, loved the taste of cod liver oil and had to be watched lest he consume too much of it. Ugh! He'd have been welcome to my share.

In 1931-32 a plague of dysentery swept through Louisiana and Mississippi. At a time when most households had no running water and everyone drank from the same dipper in the water bucket, infections were hard to control. As antibiotics were not yet discovered, drinking a juice made by boiling blackberry roots was beneficial in stopping the incessant diarrhea.

Richard had three sisters who became nurses, one of whom started giving him vitamins as she thought he was too skinny. One of the vitamins – possibly B-12 – caused him to break out in a rash so his sister took him off the vitamins, but Richard hid them in his room. When he was going to have a test at school, he'd take a vitamin pill which would result in a rash that got him sent back home by the teacher.

GAMES, OTHER ACTIVITIES AND PETS

Richard says all boys knew how to build things like stilts (also called tom walkers), slingshots, bird traps and zoozoos. A zoozoo is a large button, like that from a coat, put on a string that is tied so as to form a circle and then attached about an inch above the end of a straight stick such as a pencil. Once you get the button spinning round and round, Richard says it was fun to stand behind other people and get it tangled in their hair. (You might know he would add a new dimension to the art of playing with a zoozoo.) Jessie says that she always wanted a zoozoo but never had one -- and neither did I.

Town ball was a game played with a ball of about a three-inch diameter made of string. A bat could be anything from a fence picket to a tree limb. Another game Richard said they enjoyed playing was called "cutting hot butter." For this game a board was placed over a log and a person would get on either end to use it somewhat like a seesaw. Once when Richard was on the board, one of his cousins purposefully knocked the board sidewise throwing Richard off and breaking his arm.

Richard found a way to get even with this cousin. The boys liked to catch bumble bees, not intending to harm them, just being boys. No male bumble bees sting and even the females of some species don't sting, but there are smaller species of this insect that do inflict stings. Richard knew which ones to avoid catching but his cousin did not. Remembering his broken arm, Richard told his cousin it would be wise to handle only the smaller bees. You can guess the rest.

There was an oak tree on the school grounds where a rope had been tied from a limb. Richard said they would take turns swinging round and round on the rope. Once a girl got too wild with her swings and hit the wall of a building, breaking her arm. After Richard had told me the foregoing stories that resulted in injuries,

Jessie said, "I shudder to think of all that happened back then that would result in lawsuits now."

During school recesses, Richard and his friends often played marbles. He says, "Kids now don't even know what a taw is." (Just in case you don't know, it is the shooting marble.) In this game, each player puts a marble in a ring drawn in the sand. When it's his turn to shoot, a player who shoots another player's marble out of the ring gets to keep the marble. At some point the school principal and teachers put a stop to this game as they were "playing for keeps." One minister also warned against playing marbles as he considered it gambling.

Girls played games such as jacks, hopscotch, jumping rope, and singing and performing "Go in and Out the Window." Both boys and girls enjoyed playing on the seesaws. Jessie says that schools don't allow seesaws now because of the chance of an injury that could result in a lawsuit. Schools and park superintendents began removing seesaws from their grounds in the 1970s and '80s when the chance of being sued became more common.

Richard says, "All country kids love to climb trees. Caroline was even better at it than her brothers." Another thing most kids enjoy is riding a bike but those in the country found it challenging in the sand beds and on the gravel roads. Richard says he was the first kid in the community to get a bike, a girl's bike which was his preference.

Roller skating was also difficult as there were no appropriate places to skate. Richard says that by the 1940s his parents had a cement front porch on which he skated, but he often ended up going off the end of the porch and into the rose bushes. Sometimes he did get to go to a skating rink in Natchitoches where he had to rent skates as only certain kinds of skates were permitted there.

When Richard's Uncle Bud came to visit, he and Richard's dad would sit on the porch with their straight chairs tilted back to rest on only two legs. The men would chew tobacco and spit the juice on the rose bushes by the porch. Having her beautiful roses besmeared like this would really upset Richard's mother, but when he was telling me about this, Jessie added, "Nicotine is good to treat blackspot on roses." Could be one reason Richard's mother was recognized for having the best roses in the community.

As a child, Richard made pets of various wild animals, keeping them in a cage only until they were "tamed." One was a squirrel that had built a nest in a valance over a door. Richard always had peanuts in his shirt pocket for the squirrel. Once when Uncle Vardell, who did not know about the squirrel, was coming through the door for a visit, the squirrel jumped down on his shoulder to check him out for peanuts. Uncle Vardell screamed and bucking like a bronc, rushed out the door trying to dislodge whatever it was on his shoulder.

Richard's mother, who was in the kitchen at the time, could not imagine what was making Uncle Vardell yell so loudly. Later when Vardell said he was going to "fix that squirrel," Richard's mother retorted, "Oh, no you're not!" Richard said this squirrel remained with him for about eighteen months before leaving, perhaps to be with its own kind.

Another time the Johnson family was visiting at Uncle Bud's house. From the story Richard told me, I would say that Uncle Bud's sons didn't always treat their pets as kindly as Richard treated his. It seems that during this visit, just for a prank Uncle Bud's boys had put the family cat in the heated oven for a time.

Before those present could be anticipating roasted feline for lunch, the oven door was opened to release the cat. Caterwauling, the animal jumped out of the oven and raced out the door, scrambling over Uncle Bud causing him to fall off the end of the porch into the flower bed. The cat probably never trusted those boys again.

We all know that "Mary had a little lamb," but this Mary I'm going to tell you about had a little lion. Mary Land was a friend of

Caroline's who lived in New Orleans. She was the "Girl Friday" to Robert Maestri who was mayor of New Orleans in 1936-1946. The lion Mary had as a pet was one of twins born to a lioness that could not raise but one of them.

Once while Mary was at work, a thief came through a window of her house intent on making away with whatever he could find to steal. As he was gathering up the sterling silver and putting it in a pillow case, the lion came padding into the kitchen. The thief was so frightened that he fled out the door, leaving the pillow case full of silver behind. (Who needs a watchdog with a lion in the house?) Someone later spotted the lion lying in the open doorway and called Mary to alert her that her pet was no longer confined.

Robert Maestri, being well-known and influential, was often quoted. One time in the 1930s when he was dining on oysters with President Franklin Roosevelt at Antoine's restaurant in New Orleans, he was said to have blurted out in his strong New Orleans accent, "How ya like dem erstuhs, Chief?"

John Lynch was another friend of Caroline's that Richard had met at an iris show in Lafayette. Often when John and Caroline would visit Mary in New Orleans, the three would go ballroom dancing. John once told Richard, "I'll bet you didn't know that Caroline liked to dance."

SCHOOL DAYS AND GRADUATION

Now back to stories of Richard's school days. Richard says that in the summer, movies were shown in the school building every Saturday night with admission only ten cents. When "Gone With the Wind" came out, the school bus took children – mostly girls – to the Live Oak Theater in Natchitoches to see the movie. The price of admission there was more.

Richard told me about two daughters of J.T. Taylor who drove a pickup truck. When something broke down -- like having a leather belt break -- they would crawl under the truck to fix it. They didn't need roadside service.

The Taylor girls loved to play basketball barefoot in the sand. Since there was no gym at Readhimer High, even school basketball tournaments were played barefoot in the sand. Richard added that the sand could be very hot in the summer. He recalls running barefoot through it jumping from one clump of weeds to the next seeking the shade there.

When Richard told me about helping himself to sugar cane and watermelons from roadside fields, I asked him if he ever got caught swiping the cane or watermelons. He had this to say, "A group of seventh grade girls, including my younger sister Louisa, were having a slumber party and wanted some sugar cane. As there was a good patch down the road a short way, about a dozen of us boys went down there in a truck. Our presence alerted the owner who came out with a rifle and raked that cane patch good. Most of the boys ran into a barbed wire fence and got scratched up but one other and I knew about the fence so we dropped down in a ditch. It got pretty warm down there with all that shooting. The teachers later heard about our escapade and hit the ceiling."

Not all of Richard's escapades occurred while he was in school as evidenced by a story he told me about something that happened when he was about twenty-one. It was on a Halloween night that he and his friends had tied things like barrels and old cook stove lids behind their vehicles to create a racket as they dragged them along the road. The noise, so frightened four mules in a pasture that they jumped the fence and were located later down in Saline Swamp. The daughter of the mules' owner was in one of the cars but Richard says she never told her dad what had frightened the mules.

Richard said the walls of the school looked like prison to him. He'd have likely had a different view of school had he been taught by Caroline as was his older sister Julia. He says, "Julia would still be in first grade if Miss Carrie had not retired from teaching. The two were kindred spirits."

Although he was not fortunate enough to have Miss Carrie as a classroom teacher, Richard did have his favorites. He told me about something that had happened back in 1919 when several members of the Bishop family went down on the train to Natchitoches for the graduation of Margaret "Maggie" Bishop. As the family was returning home, the railroad bridge over the Red River collapsed just after the train crossed over it, the tracks having been undermined by flood waters. Maggie's sister Audrey Bishop later became one of Richard's favorite teachers.

Although Richard did not like formal schooling, I've never known a person who is more knowledgeable on such a great variety of subjects. Since he had a very keen interest in learning, I'm wondering if classes for the gifted and talented like those offered in the schools today could have held his interest by challenging his active mind.

Richard had wanted to join the army while he was still in high school but his dad said he had to finish school and then he could do what he wanted. When he went to enlist after graduation he was told, "We're trying to get rid of people, not hire more." Later he would have liked to join up for the Korean War but at the time he had a cast on an injured foot.

On high school graduation night, Richard and some of his friends borrowed a pickup truck to go for a joy ride. Some were in the truck cab while others were riding in the truck bed. At one point the driver went too fast down a rough hill, throwing out all those in the back of the truck. Richard says while he didn't have a scratch on him, he was covered with the blood of a girl who had hit concrete with her head. Those injured treated their wounds with iodine and such but none went to the doctor at the time. However, Richard said that the girl with the head injury was still having some problems several months later.

Another incident that Richard told me about happened one time when he and some others were shooting off fireworks. One boy was atop a wooden water tower in Saline that caught fire when Roman candles were set off. The boys had to form a bucket brigade to carry water up to extinguish the fire.

Spending one year in college at Southwestern in Lafayette Richard says could sometimes be fun but it was still like prison to him. He did enjoy biology, especially in the lab doing such things as dissecting frogs and grasshoppers or slicing earthworms. His lab partner was a girl who took less pleasure in such projects.

CHORES, SQUARE DANCING AND PARTIES

One of Richard's first jobs was helping his dad cut crossties after school. He says, "When you got old enough to see your dad over the log, you were old enough to handle a saw." Even Jessie and her mother used a crosscut saw to cut their firewood.

At this time, all children were required to help with work around the home. During rainy weather or if he and his siblings appeared bored, Richard said they would be sent to the barn to perform such tasks as picking peanuts off the vines. He says he quickly learned not to seem bored or he'd be put to work. Even when spending the night with others, children were expected to help with the chores there.

Richard told me, "We found plenty of things to do and most of the time they were legal." He said that his friends always wanted to go to Jonesboro-Hodge after school but he was "content right here."

Once it was storming when Richard and his dad were hauling slabs from crossties to use under the wash pot. Just as his dad pulled the mules under the shed still hitched to the wagon, lightning struck a hickory tree about 15 feet from where Richard was standing in the doorway. He was knocked through the doorway flat on his back, momentarily blinded and deafened. The mules which were several feet from Richard were both knocked to the ground but his dad was able to get them back on their feet shortly. Later Richard told his father, "You must have thought more of the mules than me." His dad's reply was, "I knew you were alive by the way you were hollering."

Richard often built bird traps that were constructed of many narrow wooden sticks placed in a conical pyramid with a figure four trigger. They were called "deadfall traps" as they usually did kill the birds when the trigger was set off. Any kind of bird was considered food for the dinner table. A dish that boys and girls made for many of the parties that Richard attended was robin pie for which as many as 10 or 12 of the birds were needed for a pie. This pie had become popular in the nineteenth century.

Richard first learned a juvenile form of square dancing while in grade school as a music teacher came two or three times a week to teach it to the students. He says, "Nobody in the community ever realized how sinful that was!" Later Princess Park in Shreveport gave free square dance lessons and classes for callers.

The teenagers enjoyed square dancing in the Readhimer High School gym every Saturday for some time. Richard says, "At first, square dancing was considered okay but then the Baptist preacher realized 'Hey, it's dancing!' and began including it in the many sermons he preached against dancing, not aware that his own daughter danced." Four girls from the Baptist Church often came to the Johnson's house to dance. Even back then the Methodists had no problem with dancing.

With the gym off limits, the square dances then moved to homes where often the wood cook stove was moved out so the kitchen could be used for a dance floor. The stove's flue was lifted up a few inches so the soot could be collected in a paper bag to keep from getting it on the floor. Richard told me, "We had a couple of pretty good fiddlers; all you needed then was a guitar picker or two and a caller."

In the 1940s dancing was enjoyed at Club 71 between Clarence and Campti or at the Blue Moon nightclub just north of Bunkie. This being the time of the Big Band era, the Tommy Dorsey Band sometimes came there to play.

When I was a teenager, we had parties in an old country school building where we played something we called ring games set to music. Actually, they were nothing but square dancing but we could not have called them such and performed them in the school building.

Many years later, when my sister and I were visiting Silver Dollar City in Branson, Missouri, we stopped to listen to a string band playing on the grounds there. Between songs, one of the musicians started talking about the old-fashioned ring games. Noticing my sister and me looking at each other and laughing so much, he pointed at us and said, "Yeah, I know, you danced and called it ring games so you could get away with it."

Games popular at this time included knocking for love, wink um, spin the bottle and kick the can. The first three were played at parties and the last one just any time kids got together. There are variable rules for playing each of these games.

When I asked Richard what were the rules of the version of kick the can that he played, he responded, "You mean it had rules?" Sometimes the expression kick the can, or the bucket, refers to the death of a person. Also it is used as a political expression meaning procrastination in solving problems, which is what the legislators in Washington are doing even as I write this.

Jessie had this to say about spin the bottle, "The spin the bottle game I remember was at a party given by my two sisters, the youngest being twelve years older than I. I probably was about four or five at the time. The bottle was spun and the guy that it pointed to when it stopped picked a girl to go walk around outside the house with, probably sneaking a kiss. Or if the bottle stopped at a girl, she chose a boy to walk with her. Me being me, I put up a fuss to get a chance to play so to have peace and quiet, they let me do it one time. One of the guys was kind enough to walk me around. Little sisters!"

Another variation of spin the bottle was to take turns spinning it. Players would sit in a circle with a bottle in the center. If the spinner was a boy, only girls would be in a circle or boys would be in the circle if the spinner was a girl. The spinner got to kiss the person the bottled pointed to when it stopped. Another version was to allow a hug in five seconds or a kiss in ten seconds. Once asked if he dated any of the girls with whom he attended high school, Richard answered, "I couldn't. I was kin to all of them."

Knocking for love, or sometimes called post office, began being played at parties in the U.S. as early as 1925. The game has been referred to in many movies since 1932. In the seventh X-Files episode, someone suggests that they should play post office or spin the bottle.

Shivaree is a custom brought over by the Scots-Irish to the U.S. Richard says, "Up until about WW II you were not considered properly wed unless you had been shivareed." Jessie says she remembers being a part of only one shivaree. The custom involved a group of young people going to the home of a couple just as they had gone to bed on their wedding night. Some would blow car horns while others marched around the house banging on pots and pans. Eventually the couple would arise and then the new bride was expected to cook something for the group. Often the intruders would not depart until two or three o'clock in the morning.

Since I didn't remember the practice of shivarees when I was growing up in West Texas, I consulted a Texas friend, Ronold Ray who is my age, about the subject. He said that it was quite common in his community and that most of the time it was done in good taste, but he remembers one time when things turned sour. He described the bride as "the meanest and wildest woman in the country." She and her bridegroom, who were in their thirties, were not about to cooperate for a shivaree. Unable to get the couple to come out of the house, one of the crowd of young people decided to enter a back room through a window. The bridegroom stuck a shotgun through another window and began firing, scattering the group like a "covey of quail."

RICHARD MEETS JESSIE

As I wrote in the first paragraph of this book, the first time Richard saw Jessie he knew he wanted to marry her. That first sighting occurred when he and a cousin were in the community cafe while Jessie was there with Arnold, another of Richard's cousins. Richard asked the cousin he was with, "Who dat?" The cousin answered, "Dat Jessie." Then Richard informed him, "That's the girl I'm going to marry," even though he'd never met her.

At one point, Arnold said to Richard, "You the money man. Play something for us on the juke box." Richard gave him a quarter and told him to go play it himself. Then the conniving Richard went to sit down in the vacated seat by Jessie to have a chance to talk with her.

The next time Richard saw Jessie, she was walking down the road and he picked her up. At the end of that evening on what Richard considers their first date, he asked her to marry him. She said she would, thinking surely he was just kidding. But Richard says, "I was serious." Only shortly before he had sworn he'd never marry but he changed his mind in a hurry when he met Jessie. Still in high school at the time, she says it was about two years before she really fell for him.

I asked Jessie if she and Richard square danced while they were dating. Her answer was, "By the time Richard and I met, there were more cars and movies so not as many parties and dances. Most of those had stopped during World War II when all the young guys went off to war. So Richard and I never attended any dances but went to movies every Saturday night in the school auditorium, mostly western ones like those starring Gene Autry. On other dates we went on picnics, swimming in the summer, and visited relatives and friends."

JESSIE'S BIRTH, FAMILY AND EARLY YEARS

Jessie Wade Fair was born January 27, 1934 to Simeon Eddie Fair – called Sim by everyone -- and Alice Edgerton Heard Fair. It was the standard custom in this area at the time to name children after the doctor who delivered them. Dr. Wade delivered Jessie and Dr. Edgerton, who was a good friend to Jessie's grandfather Heard, had delivered her mother.

Jessie's siblings in the order of their birth were Lance B. Fair who died in July of 2014 at the age of 99, Dovie Fair Davis who died in 1982, and Violet Fair Smith who died at the age of 92 in 2013. Jessie was twelve years younger than Violet. The family lived on a farm five miles northwest of Saline in Bienville Parish.

Jessie's maternal grandparents were Stephen Jesse Heard, who was the youngest of sixteen children, and Nancy Mathilda Goocher Heard. Jessie says, "I'm related to all the Heards in Louisiana and any other place." Her paternal grandparents were Frances Marion Fair and Dora Sophronia Fields Fair.

Jessie's mother loved to fish but Jessie herself never learned to like it. As a child, when the two would be fishing she'd constantly ask, "When can we go home?" She considered the creek could be better utilized for swimming. Between the ages of about six and nine when she was learning to swim and gain confidence in her ability, she had the protection of a one-gallon syrup bucket tied under each arm. Apparently floaties had not become available then.

Two of Jessie's aunts, Jessie and Shirley Heard, were both teachers. As did other teachers through the years, they once lived with Caroline and Virginia at Briarwood and walked to school. At the time the Dormon sisters were living in an old house which was on the site of what is now the Visitors Center.

In the summer of 1951 Jessie lived with her sister in St. Paul, Minnesota and worked as a file clerk in an office there. During this summer, Richard wrote her almost every day but she did not write him that often as she was not as fully smitten as was he. After graduation from high school, Jessie went to Shreveport to work as a dental assistant.

Jessie had attended Saline High School. Back then, a school term did not begin until late September in order for children to help in harvesting the cotton crops. Jessie says, "We all picked cotton to help buy our school clothes." With no air conditioning in those days, starting in early August as most schools do now would have made for great discomfort in the sweltering heat of the classrooms. Might as well be in the cotton patch.

ENGAGEMENT AND WEDDING OF RICHARD AND JESSIE

Jessie first met Caroline when she and Richard started dating. She says, "I had known of her all my life. Of course, everyone in the area knew of Miss Caroline and Miss Virginia." In describing Caroline, Jessie added, "She was tall, frail, but my goodness, she had energy! She considered herself very unattractive, but she wasn't. She had a beautiful smile."

Richard told me that guys always took their fiancées to see Caroline and Virginia to get their approval. When they became engaged, Richard took Jessie to get their opinion of his choice for a bride. Both soon said, "That's the girl you've got to marry!" Richard says, "That made me feel real good."

Richard says that their ready acceptance of Jessie was because of their having known her two aunts who lived with them for a while. But Jessie chided him by saying, "It was because of my personality." Although spoken in jest, those of us who know her well realize there probably was much truth in her statement.

Richard and Jessie were married March 1, 1953 in the Briarwood Baptist Church in Readhimer, Louisiana with Rev. L.P. Moreland, a Methodist minister, performing the ceremony. Although many couples at this time simply went to a Justice of Peace, Richard and Jessie chose a church wedding because he was related to most everyone in Readhimer.

Richard and his sister Louisa had beautifully decorated the church with smilax, peach blossoms and daffodils. Jessie's mother walked her down the aisle to give her away as her dad had died from a logging accident when she was only four. When the two of them came in view walking down the aisle, a very young nephew of Richard's, Mozelle's son Wayne, loudly announced for all in the crowded church to hear, "There's Jessie!"

wedding photo of Richard and Jessie

reception

Following the ceremony, Richard and Jessie headed south by car with the first stop being at Lafayette, Louisiana where Richard wanted to show Jessie his special haunts that he enjoyed while attending college there. Then it was on to St. Martinville and Avery

Island before heading home to return to work. Years later when their children were preteens, Richard and Jessie took them on much this same route, making sure that it was a time that the masses of azaleas would be in full bloom in Lafayette.

EMPLOYMENT FOR RICHARD

When Richard was a teenager, he made handles for things like hatchets, hoes and axes. He and his dad would go to nearby Choctaw Creek to select straight hickory trees to harvest for the handles. Jessie says she surely misses handles like those when their implements get broken now as such quality can't be found in the stores.

When Richard was a senior in high school he used a broad-axe to cut crossties to earn money to go to college at Southwestern in Lafayette, now ULL. When sills or other boards were needed for a new building in the area, he used the broad-axe and a froe. A mallet made from a dogwood tree was used to drive the froe into the cypress from which boards were made. His broad-axe and froe are now hanging on the back porch of the log house at Briarwood.

After college, Richard spent three years building bridges for the Highway Department. His next job was in Shreveport where he worked for Foremost Dairy for ten years as foreman of the ice cream department. After that, he worked 30 years for Martin Timber Company near Castor, Louisiana as head electrician. Richard says, "This was not an old-fashioned sawmill but all electric. It was my job to keep it running."

During his time at Martin Timber Company, people would often bring in plants for Richard to identify. He says, "Some seemed to think that I was a college professor, but I was a long way from that."

In an article in the Piney Woods Journal, Mary K. Hamner penned, "Richard is the historian and Jessie is the plant specialist, spouting scientific names like a second language." Yes, Richard can entertain for hours with historic stories of Briarwood, early times, Indian lore, you name it, but he is also an excellent botanist along with Jessie.

Richard built the house that he and Jessie lived in from 1959 until 1977 when they moved to the Sudie Lawton Headquarters Building at Briarwood. An excellent electrician, he wired the Visitor's Center which he and a brother-in-law of Jessie's had built. Although he did not build either the educational building or the pavilion, he did draw up the blueprints for each.

Once when Richard was installing rest rooms in one of the buildings, there was an interruption by an act of nature. As he entered the men's rest room one day to continue his work, he discovered that a wren had built a nest on the wash basin. Construction there was discontinued until the nesting cycle was completed and the baby birds fledged. Richard says, "It was as Miss Carrie would have wanted it."

This story reminds me of a like incident that occurred when my youngest son was building his house. A wren had entered through an open window to nest in one of the bedrooms, arresting work there until the bird had completed her nesting duties. And like Richard says of Miss Carrie, this was just as I wanted it. This son's degree is in wildlife biology, so you can expect him to be protective of the wildlife. And another of my sons claims that, rather than hunt animals, he's on their side.

After retirement from Martin Timber Company, Richard stopped by the office one day to chat with the girls there. They told him that things were dull around the office without him, asking, "Won't you come back?" But Richard had not been the only practical joker to keep things lively at the company. Once another worker put a dead rattlesnake on the stairs hoping for what he considered would be a startled reaction. It was. When the mill foreman spotted the snake, he quickly began beating it with a stick thinking it was alive.

A bent for practical joking must have run in the Johnson family as Richard's younger brother, Leslie, had an artificial snake he used to create a little excitement from time to time. Once when he spotted a black man backed up under a shelf performing some job, Leslie flipped the snake on him. Startled and angry, the man came out from under the shelf to drag Leslie into a back room. When I asked

Richard what happened then, he answered, "Let me put it this way: Leslie didn't drop any more snakes on that man!"

I first met Leslie when I was attending a workshop entitled "Becoming an Outdoors Woman" where he was an instructor in canoeing. Because the creek was at flood stage, we canoed in the swimming pool. Had I known then Leslie's reputation for staging practical jokes, I might have been hesitant to get in a canoe with him!

After Caroline's death while Richard still held a full-time job elsewhere, Jessie worked almost daily at Briarwood and Richard worked with her there on weekends. He had constructed a flagged route of about a mile through the woods for Jessie to follow to get to her work site.

BIRTH AND EARLY YEARS OF RICK AND ANN

Richard (Rick) Michael was born August 22, 1954 to Richard and Jessie. Richard had a boat and motor borrowed, planning to go fishing that day. He says, "Rick messed up a good fishing trip."

Rick and Ann as children

On June 21, 1980 Rick married Denise Loretta Valdez, who was born June 27, 1959. Their children are Vivien Lea, born April 28, 1983 and Richard David, born November 17, 1989. Vivien's children are David and Rebecca.

Caroline Ann was born March 20, 1956. When Richard took Jessie to the hospital for the birth, he then had to take Rick to the home of friends. By the time he got back to the hospital, Ann had already arrived.

Ann married Anthony (Tony) Eugene Hough on July 7, 1976. He was born January 18, 1955. Their children are Jacob Adam, born October 19, 1980; Daniel Luke, born March 15, 1983; and Joshua Mark, born June 23, 1986.

In an interview with Rick he told me that his earliest memories of Miss Carrie were going with his parents to visit on Sunday afternoons. As they would be sitting on her front porch chatting, she would suddenly pause in mid-sentence and ask, "Did you hear that?" It would be some bird that she recognized by its song. Not only could she identify birds by their songs, she could give a good vocal imitation of the songs of many birds common at Briarwood.

Rick spoke of the old hothouse behind the log cabin that he remembers as having a strong musky smell. It's interesting that many of our strongest memories of things from long ago are of odors, whether they were pleasant or repugnant.

As children, Rick and Ann would get off the school bus in the fall and go to Briarwood to pick up white oak acorns for which Caroline paid them a dollar for each large grocery bag of them collected. She didn't want too many seedlings from the acorns to come up where not wanted among her other plants.

Rick told me he sometimes did odd jobs for Caroline such as climbing up trees to remove dead or ill-placed limbs. He and Ann also earned money picking up pecans down below their grandfather's house. He says, "We got enough money to buy a few firecrackers."

Their parents took Ann and Rick on many trips to further their education; they wanted them to experience all the wonders of nature. Rick remembers one fall trip for which they got up around 2:00 to 3:00 a.m. to leave home early. The children slept on the way, awakening in South Louisiana in the Rockefeller Wildlife Refuge. At sunrise many ducks and other water birds were sighted swimming in the water. Rick remembers the whir and beating of the wings when something would alarm the birds, causing them to lift off the water in huge flocks.

Another time Ann and Rick were awakened while it was still dark to go outside to experience meteor showers. Rick says, "The meteors looked like rain coming down through the atmosphere. Although this was in the late 1960s, I've never seen another to match it." Even though it was with great reluctance that he and Ann got out of bed that morning, they both now recognize it as another of the times their parents were intent on creating wonderful memories for their children.

Rick says he had many cousins to lead astray. For instance, he and several of them would run though a field of grass burs to get to the pond for a swim. He knew how to jump from clump to clump of the grass to avoid the burs but didn't share this bit of strategy with his cousins so they received many painful pricks. Shame, shame on you, Rick! He told me, "My feet were like shoe leather. I tried not to wear shoes in warm weather except to go to Jonesboro or to church."

Besides having a great sense of humor, Rick says his dad has a remarkable understanding of human nature and how people will react in a given situation. The following story illustrates this latter trait. Rick and some of the other boys wanted to go swimming in Saline Creek at the Cloud Crossing but the creek was at flood stage. Knowing the water to be deceptive, Richard said, "Don't know boys, have to check it out to see if it's safe. Hold up, boys, hold up!"

Richard then took a rope, tied it to one of the older boys and permitted him to enter the water. The current swept him up, down, up, down, up, down and then flung him against the shore. After Richard pulled him to safety he announced to the boys, "I knew this would happen but I had to show you or one of you would have jumped in and probably drowned."

Remembering that Richard had told me about skinny-dipping in his youth, I asked Rick if he had engaged in such a practice. His answer, "Yes. It's a common practice for boys out in these woods."

In an interview, I asked Rick how he liked school. He said, "School was not too bad except when the windows were open and I'd get to thinking about hunting, fishing and swimming. You can see the apple didn't fall far from the tree."

Rick's high school graduation

Rick took part in Boy Scouts for about five years, attaining the rank of Star Scout with his dad as scoutmaster. He says, "Many young people in this community owe much to Dad for his work with the scouts."

Having a love of nature ingrained in him, Rick especially enjoyed the camping trips in both Louisiana and Arkansas as a scout. Two of his favorite places in Arkansas were Crooked Creek near Albert Pike and Little Missouri Falls. In later years Rick became a cubmaster for his own son and for a brief time served as a scoutmaster while living in Texas.

On August 1, 1972 Rick joined the Marine Corp. (Although Rick had grown up answering to the name of Mickey that changed to Rick in the Marine Corp.) He was stationed in California part of the time and also on the East Coast at other times. Following in his parents' footsteps, after Rick had children he made certain that they had the opportunity to experience all of the things of interest nearby wherever they were living.

After a 22- year career in the Marine Corp, Rick retired and moved to Irving, Texas where he taught for nineteen years in the high school there as head of the Junior ROTC, a leadership program.

After retiring for a second time, he returned to Louisiana and in 2013 he took over as curator of Briarwood. Those of us who love Briarwood hope he doesn't plan a third retirement for many, many years!

Rick spoke of the numerous visitors that come to Briarwood. He said that when any from the forestry industry visited, he could hear the coins clicking in their heads when they realized the value of the trees growing there. But Rick added emphatically, "They'll never be sold." Concerning 'Grandpappy,' the longleaf pine at Briarwood that is over 300 years old, Carolina once penned in a letter, "Those pesky lumbermen were here again today trying to convince me that Grandpappy wouldn't outlive me. But I know better – my very soul lives in that beautiful old gnarled and weather-beaten tree."

Grandpappy

In speaking of his daughter Vivien, Rick said that she was a member of the marching band while in high school. In football season, her parents stayed up many late nights awaiting the return of the school bus carrying the band. He said, "She still owes us big time for all those hours of sleep we lost."

When Rick told me of some of the mischievous things his son David did as a child, I said that he sounded just like my oldest son. If you met Ronnie, you'd immediately like him as he's personality plus. But he was a handful to raise as he was always full of mischief. (What do I mean WAS?)

Rick said, "If my son had been born first, I probably would have had only one child!" I'm slow to learn; I had two more after Ronnie but he holds the record for getting into mischief. That Ronnie realized he'd been a challenge for his parents as a youngster was evidenced by a printed Mother's Day card he sent me a few years ago. It read, "Congratulations, you still look great! You'd think that any woman who had raised a son like me would look like hell by now." Strong language but it illustrates a point.

When I interviewed Ann she spoke of visits with Miss Carrie when she was but a toddler. She said, "I always loved listening to her stories of Indians, birds, other animals, plants, all about nature." Jessie says that having their children associated with Caroline opened up a whole new world for them.

Ann says she enjoyed walking in the woods and exploring them. Once on such a walk she discovered a whorled pogonia orchid which she took back to show Caroline, who was excited to have Ann share it with her.

Another time Ann and her mother accompanied Caroline to the woods to pick sandhill grapes. Noticing Caroline standing in a bed of plants, Jessie asked, "Miss Carrie, you're not allergic to poison ivy? Isn't that what you're standing in?" Looking down, Caroline informed her that it was fragrant sumac which has trifoliate leaves similar to poison ivy but does not cause allergies.

<u>Jessie and Ann with mallard ducklings</u>

Because Caroline was thin, Ann told me that to her "she always looked hungry." Although Caroline was a good cook herself, Dosia, the daughter of Nora Patterson, usually did the cooking for her. Ann remembers a special delicious shortcake that Miss Carrie cooked which was more like a pie crust.

Ann also spoke of some of the same trips that Rick had told me about. She says they got "a good Louisiana education" learning about the plants and wildlife. Their first trips were in Louisiana with later ones being camping trips to Arkansas. Some of the things that really captured her attention there were lady-slipper orchids, quartz, and fossils in the rocks in the area of Richland Creek.

It was on trips to Arkansas that Ann got an early interest in the mountains. The Johnson family had first explored Louisiana, then Arkansas and on to the west. Ann says, "We graduated from Arkansas to the western states." She now lives in Wyoming surrounded by the mountains she loves.

Ann told me that she likes to hunt rabbits and deer for which Richard had given her a hunting rifle when she was a quite young adult. He says, "She kills more deer each season than anyone else." When I was interviewing her during deer season she said, "I've killed one and a half this year." One deer was so small that she thought it would be misleading to say she had harvested two.

Once when Ann and Rick were teenagers, Richard had made them a cannon using a 4 ½ inch pipe on which he closed one end. A small amount of gasoline could be poured in the mouth of the pipe and touched off with a lighted match, creating an explosion that could be heard for over a mile. After checking it out with the town marshal to get his approval, it was taken to Saline Lake to set it off. The marshal was all for letting kids have fun as long as it didn't destroy anything.

Ann's high school graduation

One time a salesman had given Richard a couple of puzzles that appeared identical; however, one could be taken apart but the other could not. Even to be able to take the former apart required a person to do so in a certain order. When her dad gave them to Ann, she took them to school and gave a brief demonstration to the teachers of taking one of them apart, but then played a trick and gave the other one to the teachers. (Think another apple didn't fall far from the tree?) Richard says, "Ann was a Johnson; she could play jokes."

Later the principal, who is now Ann's father-in-law, told her to take the puzzle home. He considered that the teachers were spending too much time in the lounge trying unsuccessfully to get the puzzle apart. I wonder if he tried it himself?

Ann is a registered nurse who now works in a hospital emergency room in Wyoming, having worked earlier in a Ruston hospital. In an interview she told me that in Ruston state troopers would bring drunks involved in auto accidents to the ER before taking them to jail, but in Wyoming she says they are taken directly to jail.

A SURPRISE FIFTIETH WEDDING ANNIVERSARY CELEBRATION

When Johnson relatives gathered at the Interpretive Center at Briarwood for their reunion in June of 2003, they surprised Richard and Jessie by making it a fiftieth wedding anniversary celebration for them. Richard's sisters, Mozelle and Louisa, gathered letters written by a number of relatives to be presented to Richard and Jessie. Following are some of those letters.

This one was written by Richard's sister Nettie:

June 2003

Dear Jessie and Rich:

There have been so many wonderful times that Tommy, Mike, Margaret and I have shared with you. I don't know where to begin.

One of the funniest that I remember was before Jesssie could drive a car but she could drive the tractor. Rich put a seat on back so she could take the kids for rides. Mike, Margaret, Mickey and Ann were riding on the back with Jessie and I in front. We were just about at the old mail box when Jessie yelled, "Mick get that snake on the fence for your collection." Mick yells,"Come on Mike." They both sail off the tractor and are about half way to the fence across the sand bed when Mike comes to screeching halt and yells, "What kind of a snake?" By the time Mick got to the fence the snake had gone.

Another real funny time was on Tommy and my 50th wedding anniversary. Jessie and Rich had worked so hard with Tommy to keep it all secret from me. But on Sunday morning of the ceremony Jessie, Tommy, Margaret and I were in the kitchen and Jessie told Tommy she thought he should propose to me in case I didn't want to marry him again. Tom wasn't too happy about that but he did it with Jessie and Margaret as witnesses. When we got to church and the minister announced that we were getting married right after church in the cemetery I almost had a heart attach.

Happy anniversary,

Love,

nettie

Following is a letter written by Mike Nork, Nettie's son:

June 14, 2003

50 years together—wow! What a great relationship. Being single, I don't have anything in the way of first-hand knowledge, but I did have the opportunity to observe my parents during their 50+ years of marriage. I know that it's not always fun & games. It must surely take a lot of work, understanding, and prayer for two people to stay so close for so long. Having shared interests, such as your mutual fascination with the natural world certainly helps, as well. Continuing on with Caroline Dormon's work benefits the rest of us, too—and not just because it gives us someplace to hold the family reunions!

Love,

Mike

Mike Nork

P.S. Have you started making plans for your 100th anniversary, yet?

Mozelle, Richard's sister wrote:

> Remembering 50 years
> with Richard and Jessie
>
> We had the pleasure of being at your wedding 50 years ago and now celebrating your 50th anniversary. There have been many laughs and a few tears in those years. God and miss Carrie brought you together - a perfect team.
>
> Rick you finally used some good sense in finding Jessie and I have loved her all these years.
>
> Jessie thank you for loving and caring for my hard headed (but lovable) big brother.
>
> May you have many more happy years together
>
> With love and best wishes,
> Mozelle

Wayne Darsey, Mozelle's son, wrote of fishing with his Uncle Richard:

Wayne Darsey

THE YEAR WAS 1956 AND I WAS 8 YEARS OLD. I HAD BEEN FISHING WITH A CANE POLE FOR 4 YEARS AND IT WAS TIME FOR ME TO GRADUATE TO A ROD AND REEL. I WAS FACINATED WITH THE OPEN FACE SPINNING REEL THAT UNCLE RICHARD USED. SO AFTER BEGGING FOR ABOUT A YEAR HE AGREED TO TEACH ME HOW TO USE IT AND THE FINER POINTS OF BASS FISHING. HE SAID "SURE" I'LL SHOW YOU HOW. "he-he-he" AND GRINNED.

SO OFF TO THE POND WE GO. I WAS SO EXCITED AND I JUST KNEW THAT I WOULD BE ABLE TO THROW THAT LURE JUST PERFECT THE FIRST TIME I TRIED. I HAD WATCHED UNCLE RICHARD DO IT SO MANY TIMES BEFORE THAT I PROBABLY KNEW HOW ALREADY. BUT TRUE TO HIS WORD UNCLE RICHARD VERY PATIENTLY SHOWED ME EXACTLY HOW TO HOLD THE ROD AND DO EVERYTHING JUST RIGHT. "OK" HE SAID, NOW YOU DO IT AND HE TURNED AROUND AND RAN. WHEN I LOOKED UP HE WAS BEHIND A LARGE TREE. UNCLE RICHARD HAS ALWAYS BEEN SAFETY CONSCIOUS.

WITH A MIGHTY SWING OF THE ROD I MADE MY FIRST CAST. THE LURE LANDED IN THE TREE . AIN'T NO FISH IN THE TREES ,

Wayne Darsey, Cont.

HE SAID. MY NEXT CAST LANDED IN THE GRASS . I CAN NOW HEAR WHAT SOUNDS LIKE GIGGLING FROM BEHIND THE TREE. AND HE SAID, GRASS BASS ARE EASY TO CATCH BUT TOUGH TO EAT. IF I WOULD FISH IN THE WATER I MIGHT CATCH SOMETHING GOOD TO EAT. "WORDS OF WISDOM FROM MY UNCLE RICHARD."

WITH THE PASSING OF TIME, I LEARNED THE SPORT AND THE ART. PASSING THIS KNOWLEDGE ON TO OTHERS IN MY LIFETIME.

MUCH OF MY PASSION AND KNOWLEDGE FOR THE OUTDOORS HAS COME FROM "UNCLE RICHARD". FOR THIS I WILL BE ETERNALLY GRATEFUL.

LOVE,
WAYNE DARSEY

P.S. A memory is a terrible thing to waste.
This is one of mine.

May 15, 2003

Another of Mozelle's sons, Glenn Darsey, had some unusual tales to tell about frogs, as well as writing about Jessie's cookies he enjoyed so much. Jessie says, "I didn't know until I read his letter what he called my Boiled Chocolate Oatmeal Cookies."

Glenn Darsey

MEMORIES OF AUNT JESSIE AND UNCLE RICHARD

Ah yes, "MEMORIES"! They are a wonderful thing. Over time, we tend to forget things, especially as we get older, but some memories stay with you for a lifetime. These are some of those memories.

FROGS IN THE WELL

I always enjoyed my trips to Saline. I could always go to my cousins house(Mickey & Ann), and we would have a grand time. I remember showing up one bright sunny day and asking my aunt Jessie.....'Where is Mickey'? The response was NOT what I was expecting. She said ' He's in the well'. AHHHHHHHHH....excuse me? She said 'yea, he is around the corner of the house, in the well'.

Always ready for an adventure, I trotted off to the other side of the house to check this out. There stood Uncle Richard over this big hole in the ground giving instructions. 'Do you see anymore? We gotta be sure we get 'em all'! I be-bopped over and asked, "What ya'll doin'? To which he replied 'Gettin frogs out the well'. Apparently, this was a common occurrence in the country to which I had never been made aware of . I looked down this big hole to find my cousin Mickey floating around in the water, catching frogs! To be honest, it kinda looked like fun but for some reason my uncle Richard would not let me get in there and catch a few myself.

As strange as that tale may seem, it did (even at that early age) cause me to consider if I wanted to drink the water at my aunt & uncles house! The image of frog pee in the water just would not go away!

DOO-DOO BALL COOKIES

Anyone with a brain that has tried some, will tell you that my Aunt Jessie is a GREAT COOK! While at her house one summer, she decided to cook us some cookies. These were the strangest looking cookies I have ever seen. They looked kinda like brownies but tasted like candy! Great combo for a kid! It turns out, they were oatmeal drop cookies, cooked in a pan. I loved them so much I got her to give me the recipe! Even as a little kid, I learned to cook those cookies. They were and are my FAVORITE cookies of all time! I love them even better than OREO'S! Go figure.

In any event, it was only later in life that I found out that Aunt Jessie was doing them wrong. She would cook them in a shallow pan, like brownies and they were "supposed" to be spooned out and dropped on wax paper, thus causing them to look like D0O-Doo balls! And that is what I have called them ever since!

— Continued

THE GLORY HOLE

On another bright sunny day I arrived at Uncle Richard & Aunt Jessie's house and found a whole new project had been started! OH BOY! A new adventure! It seems that my uncle had decided to dig a GREAT BIG HOLE in the Pine thicket just above his house. Folks, I am not talking a hole you dig with a shovel, he dug this sucker with a TRACTOR! This hole was HUGE, and when you are a young kid, it just looked bigger!

I was told this hole was for storing potatoes. (Go figure!) The thing is, I don't remember ever seeing any potatoes in it. What I remember is, the thing collected a lot of water in the bottom and guess what we had then??? That's right!   FROGS!!!!!   Again with the FROGS!!!

That hole became the greatest playground for us kids I ever saw! Every time we came to visit, we made a bee-line for that glory hole to catch, shoot, spit on and throw dirt clods at the FROGS in that hole. It was great fun!

I bet, if you were to go up in that pine thicket today, you might still find that big old hole where we spent so many days having fun. Thanks Uncle Richard! That was a great hole!

There are SO many stories about my time at Uncle Richard & Aunt Jessie's house that I could go on for days, but if I was to do that, I would be obligated to put it in book form and distribute it nationwide! So rather than embarrass you, I will stop here. I want you to know that a major part of my youth was spent around you and other than my parents, the two of you had a great deal to do with making me the kind of person I have become (good or bad!). I want to thank you for all the great times you gave us when we came to see you. I also want to thank you for putting up with all the "STUFF" we did while we were there! Especially, the things that EDDIE did. It is a wonder he survived.

Thanks to both of you for all you have done for me over the years! Your the greatest and I love you dearly!

Your Nephew,

Glenn Darsey

This next letter was written by Bret Fremming, the son of Richard's sister, Louisa:

From: Brett Fremming
To: bdfre@earthlink.net
Date: 6/4/2003 10:28:57 PM
Subject: memories

I remember three particularly fond memories from my childhood visits to your farm. First, was going into your fields with you to pick REAL watermelons and digging for potatoes. At the time, I thought that was the greatest treasure hunt going and how lucky you where to be able to just go outside and pick up your own favorite food groups and eat them immediately.

Secondly, I remember the raft you built for the pond with four 55 gallon drums and a wooden frame. I recall playing Tom Sawyer. This was my first time at sea without parental supervision and the sense of independence was terrific.

But my favorite thing was shooting the carbide cannon. That was so much fun that I'm suprised it was legal (or was it ?).

More importantly is the example of love, devotion, and faithfulness that your 50 years of marriage has shown to us youngsters. Long term relationships are rare these days and you should be proud (especially since you two have yet to ever have a disagreement).

Go for another 50.

Bret Fremming

Richard's sister, Isobel, wrote this one:

Jessie & Rick has meant so much to me & mine over the 50 years, that I can hardly begin to say it all!

December 1979, Christmas, after Frank died in Nov. 1979, Patty came to Briarwood and spent the holiday with them. They were so wonderful to us and made our grief much more bearable! We saved the package of wood splinters from them for years!

Never will I forget April 22, 1995, when I was driving down # on my way to Briarwood to go to the Strange Cemetery service — Suddenly, I was in several feet of water & my car was floating down a branch of the river! The Jonesboro Rescue Squad & 2 men from the La Road Dept. got me out & call Richard. He braved more flooded roads to the Jonesboro Hospital to see me & take me to Briarwood. Jessie was so wonderful too. She washed & dried my clothes and pictures, billfold, purse & other things!

As always over the years they've had us there for family reunions and other times. All of my family feel they have been so wonderful over the years, and we love them!

Ibb

Steven Hickory, Isobel's son, wrote of having fond memories of visiting Richard and Jessie as a young child:

> ## 50 Years of Love
> 6/03
>
> I've always admired Uncle Richard and Aunt Jesse since I was a young child taking summer vacations on the farm. I can't think of two nicer people who have always seemed just right for each other. I remember how hard both of them worked around their house. Even though they didn't seem to have much money, they still managed to live a wonderful life. They both seemed to enjoy reading National Geographic and just about anything with interesting content about nature. I also liked the way they were involved with their kids. They'd take all of us to the fire tower or to the creek to swim. After all these years they are just the same and just as wonderful. Congratulations on 50 years of love.
>
> Steven Hickory

Isobel's daughter, Shari, spoke of the inspiration Richard and Jessie had been to her and husband Dave:

> Dear Aunt Jessie and Uncle Richard, —
>
> 6/1/03
>
> You have been a wonderful encouragement to Dave and I for the last 30 years of our marriage!
>
> When I think of you both I am filled with joy and a cozy warmth, the same feelings I have when I think of love and home. You have been such an inspiration to us!
>
> Congratulations, both of you!
> And may this card convey
> All kinds of happy wishes
> Sent in honor of your day!
>
> Wow! Congratulations on 50 years together!!
>
> Love,
> Shari & Dave

In Rick's letter he spoke of many cherished memories from childhood of family time together:

Memories with Mom & Dad

Asking me to write about my most favorite memory of my parents is one of the most difficult yet enjoyable projects I've ever had to do. There's no problem thinking of wonderful moments, only trying to think of one that's above the rest.

Which do I choose? The memory of family time together obviously tops the list. Camping in Arkansas along Crooked Creek, or maybe the times spent along Richland Creek gathering fossils and swimming in the hole at the base of "Falling Water Falls". What about the long walk up State House Mountain, just to see the rock in the meadow, or up Crystal Mountain to explore and find some of the most beautiful quartz in the world? How about swimming in Crooked Creek, Little Missouri River, or any other body of water that was deep enough to get in? Say Arkansas and a flood of wonderful times come back to me.

Louisiana will do the same too though – trips to south Louisiana – Avery Island, New Iberia, Iberville, St. Martinsville, the Evangeline Oak of Longfellow Fame, the Acadian village, the smell of fresh baked French bread from a bakery somewhere in that area. Seeing the coast line with all its areas to explore, the awesome site of an early winter morning after what seemed like an all night trip to the Rockefeller Wildlife Refuge. The noise of what seemed like and very likely were millions of water foul was amazing! Then to see them take off in a huge swarm whirling through the sky in a dazzling blur of motion and sound –

These trips were wonderful; they were filled with fun, all the while learning about the world around us. . . . . .

Trips, though special, weren't the only times for special memories – home was full of them too. Holidays hold many memories, coloring eggs at Easter, Thanksgiving with family coming over for dinner, and Christmas with the whole gang going out to get that "perfect" Christmas tree, for which the hunt had been going on for months. It took years for me to get over not having a "natural tree" for Christmas – really haven't still. Though holidays brought additional things, daily life with Mon and Dad was very special too. Things like – being wakened at something like 2:00 a.m. to see a meteor shower that looked like rain – Dad teaching me to fly fish, learning to cook by watching Mom – and they both made sure I learned to wash the dishes!

I guess what I'm trying to say is that my whole life with Mom and Dad has been a special memory, one that I will cherish all of my days.

Rick's daughter, Vivien, spoke of her grandfather as a wonderful storyteller and her grandmother as a great cook:

> You know, it took me a long time to think of a fond memory of my grandparents; not because I don't have any, but because there are so many. From fishing, to canoeing, watching raccoons, and learning about nature, the list could go on, so as you can see it took me a while to think of the best memories that I will always cherish.
>
> For Grandpa, it would be the late winter nights, the whole family sitting around the fire place listening to one of his wonderful stories. The tales are endless and you will never hear the same one twice. There are stories of family, friends new and old, animals, and the crazy Boy Scout troops.
>
> My memories of Grandma however, are in the kitchen. I enjoyed spending time with her, learning how to cook, but mostly getting in the way. The meals always turn out perfect, and a lot better than any restaurant could do. I've always wanted to be able to cook as well as she does, but according to my newly wed husband, that is a task I have yet to develop.
>
> These are the memories I shall cherish forever, and the stories I will share with my husband, my children, and my children's children.
>
> ~Love always,
> Vivien Lea Johnson-Lavely
>
> Happy 50

Rick's son, David, related a few of his fond memories of trips with his grandparents:

> My most memorable moments with my grandparents were spent camping at Crooked Creek. On one trip, crystals were on our agenda. We walked up Crystal Mountain, stopping every once in awhile to pick up a pretty looking crystal. Grandpa told us about his history with the mountain, as we slowly reached the top.
>
> I must say they had taken me to the most wonderful place, almost as wonderful as them!
>
> Another time we went to Oklahoma to go fossil hunting. Grandma and Grandpa had more knowledge on the subject than a super computer's memory could ever hold! As I found different types of fossils they answered my questions one by one about these fascinating objects. They even showed me the living descendants of one of the fossils!
>
> Grandma and grandpa have always been an inspiration to me. I've always wanted to be as smart as them. They shall always be great people in my mind, and heart.
>
> <u>Happy 50th anniversary</u>
>
> *David Johnson*

Jessie says this is only a sampling of the love expressed by so many at the anniversary celebration. "Those letters and the love extended by others in person really made us realize how much we are loved. That love is returned by us."

Fiftieth wedding anniversary celebration for Richard and Jessie

Of Richard's eight siblings only Isobel, Louisa and Leslie are now surviving. The dates of death for the others are: Willie Mae, March 9, 1927; Jerry Willard, December 19, 1995; Julia, September 2, 2000; Nettie, January 7, 2006; and Mozelle, September 16, 2012.

ORIGINATION OF BRIARWOOD

Briarwood was originally a plantation established on the Edisto River in South Carolina and inherited by Caroline's grandmother. It was moved to its present location in 1859. For the move, the slaves and some of the white men came by wagon while the women and children came by boat. Dr. Benjamin S. Sweat, the grandfather of Caroline and Virginia, was on the trip accompanied by his son James Lawrence. When the Civil War broke out, James joined the Confederate Army. He was later accidentally shot by his own men while he was on surveillance duty and died of gangrene.

On arrival at their destination, the slaves asked Dr. Sweat for a place to bury two slaves who had died on the trip to Louisiana. A spot on the original acreage of Briarwood, located on what is now called "Brick Yard Branch," was provided for the burial. A few years ago, a black man from California came to Briarwood wanting to visit the graves of his ancestors. Richard took him to the small cemetery and each shared with the other what he knew about its history.

There are now numerous buildings on Briarwood that have been constructed through the years. Caroline and Virginia first lived in a house that was built circa 1916 on what became the site of the present Visitors Center. They later moved to the log cabin which was built in 1950. The Sudie Lawton Headquarters Building was constructed in 1977, the Writer's Cabin restored in 1979, the Visitors Center in 1980, the Interpretive/Education Building in 1999 and the pavilion a couple of years later.

The first writer's cabin which was built circa 1936, was named the "Three Pines Cabin" for the three ancient pines growing in front of it. The cabin had so deteriorated that it had to be reconstructed in 1979. By that time one of the pine trees had fallen so the name was changed to Writer's Cabin.

Richard and Rick working on reconstruction of chimney of Writer's Cabin

Before World War I a young man, Clarence Lindsey, lived in a cabin near Briarwood and looked after the place for Caroline and Virginia. He grew watermelons and cotton on a hillside at Briarwood south of the Headquarters Building site. In about 1937 he planted slash pine on that hillside.

A talented do-it-your-selfer, Richard built all the cabinets and some of the other furniture in the Headquarters Building which became the home for Richard and Jessie. Richard says, "We had to wait 15 years for this, but you have to wait for the good things in life."

Richard on tractor building Headquarters Building

In the Caroline Dormon Nature Preserve Newsletter for January 1, 2011, Richard wrote the following:

> On the first Saturday of December, I decided to review the new land the foundation has purchased on the southwest side of Briarwood. As I walked along the old Sparta Road memories from my youth came flooding back! A depression fifty feet in diameter is still there. We children envisioned it to be a buffalo wallow but I doubt that, even though the old road is an ancient buffalo trail. On a sloping hillside was a large field now eroded with years of neglect which nature is slowly covering with yaupons, wax myrtles and pines. We will help by now planting the acorns of red oak, black oak and cow oak to grow into trees for the future. All of these oaks produce preferred wildlife food. It is not uncommon for one or more deer to bolt from this area as I pass by on the electric cart.
>
> Off to my left is a flat area that once was a rice field back in the 1860s. It continued across Mr. Readhimer's mill creek to form a field that could be flooded when needed for the rice. The water came from a spring-fed pond to our west. Now I have reached the creek crossing where you can walk over the metal footbridge or use the ford to wade across. Here the

water has swept clear a two-foot deep hole so clear that you can see the minnows on the bottom. A stone's throw upstream the water rushes through a break in what was a large dam. In the 1800s this impoundment powered the grist mill. The old water wheel still leaned against the dam when I was a boy. We are dedicated to preserving this newly purchased land and the adjoining 24 acres donated by Thomas Nichols and his sister, Dorothy N. Hughes.

Then in the newsletter dated January 1, 2013 Richard wrote of more land added to the current acreage of Briarwood:

> You will be happy to know the thirty acres of Briarwood lost during the Great Depression are back in the fold! We are so grateful to everyone who made this possible by giving donations toward this purchase. Thank you for your generosity! These donations and money from the Foundations discretionary fund allowed it to be paid in full. We are so grateful to Tommy Murchison for the many hours he spent doing legal work that would mean a clear deed to the property. I am also grateful for the hours donated by Julie Callihan, the Foundation's president; James Durham, treasurer; and Alanna McCormic, all so willing to help, never complaining. Again I would like to thank Judy and Sam Bayliss for their decision to sell and their daughter Alicia for her contributions. I wished and prayed but never expected this to happen in my lifetime. We will hold an open house later for you to come see the latest miracle.

In the same newsletter, Richard wrote of the home place of Caroline's grandparents, Dr. Benjamin S. Sweat and his wife Harriet Trotti, who settled there in the winter of 1859-1860. He described the house as having wrap-around porches, but that no photographs of it are now available. He said, "We found hundreds of pieces of locally-made brick and a rather large sandstone rock when we cleaned the site off this winter and even uncovered a yellow-spotted salamander in hibernation."

COTTON PICKING, INDIAN LORE AND MISCELLANY

In an interview, I asked Richard if he ever picked cotton and he answered, "Absolutely. Just something that had to be done." In the part of Texas where I lived as a child, cotton was usually harvested by pulling boll and all as an infestation of boll weevils in that area caused the cotton to be difficult to remove from the boll. Now farmers simply spray to defoliate the cotton stalks and then use mechanical means of harvest. As a Texas friend of mine says, "Cotton picking has now faded into the Texas sunset." When I asked Richard if he ever pulled bolls his answer was, "The only bolls I pulled were to throw at my sisters."

Richard says he recently viewed a TV program about Senator Adam Clayton Powell. Part of the segment from the 1940s concerned a black woman who disagreed with many things the senator was doing even though he was a Civil Rights activist. She told him that he couldn't understand their problems as he'd never picked cotton.

On the subject of picking cotton, Richard told me that his sister Nettie once had a top service job in Wheaton, Maryland where she and her husband had moved after their children were grown. One day her supervisor complained to her, "Mrs. Nork, we're having a problem with these black workers. All they know how to do is pick cotton." Nettie had a quick response to set him straight, "Now wait a minute, in the South you didn't have to be black to pick cotton. Everybody did." Richard said that the Yankees seemed to think that picking cotton put you in a lower class.

Richard says that at one time he baby-sat a little one-year-old redheaded girl while her parents picked cotton. Many years later she told her husband about Richard riding her on his bicycle.

Richard is a pleasure to interview because of his many interesting stories. But if I seem to jump from subject to subject in this book, it's because the things he tells me are not necessarily in sequence, he just tells his fascinating stories as they pop into his mind. I'm sometimes stumped as to where they should appear in this book. But who cares?

This next story is about a visit Richard's sister Isobel made to Richard and Jessie on April 22, 1995 during what proved to be a ten-inch rainstorm. With the rain coming down hard between Quitman and Friendship, Isobel's car floated off the road. She crawled out the car window and clung to a tree covered with poison ivy, never having learned to swim. Sometime later the Jonesboro Rescue Squad and two workmen for the highway department found her and took her to the hospital where she was treated for hypothermia. And poison ivy rash!

In June of this same year, Richard's sister Nettie and her husband Tommy Nork had celebrated their fiftieth wedding anniversary at the Strange United Methodist Church. Following is a photograph of the many relatives who attended.

50th wedding anniversary for Richard's sister Nettie & husband Tommy Nork

Knowing that Caroline had many stories about Indians to tell, I wondered if Richard got his interest in Indian lore from her. He says as early as he can remember he was always sympathetic with the Indians. He told me. "I just wished I had been born an Indian. I had an ulterior motive as I thought Indians didn't have to go to school."

He was about six years of age when he first realized that Caroline had a lot of information about Indians that he could gain from her. Caroline had a good relationship with the Indians; although most of her friends called her Miss Carrie, the Indians' name for her was White Flower. Attesting to his fascination with Indian lore, Richard has an original of the book *Myths of the Cherokee* written by James Mooney and published in 1900.

Richard's dad rented 40 acres to farm that had the Sparta Road running through it. This road had been used by the earliest men coming to this part of the country, by buffalo and by the Caddo Indians. Richard found many artifacts along this road and took them to show Caroline.

In the mid-1930s, after the De Soto Commission was formed to study and pinpoint the route Hernando De Soto took through the southern states, the chairman for the commission, Dr. John R. Swanton, came to Briarwood to visit Caroline Dormon, who was the only woman appointed to the De Soto Commission. Dr. Swanton was also head of the Bureau for Ethnology, Smithsonian Institution in Washington, D.C.

He and Caroline were sitting on the front porch when Richard approached them to show Miss Carrie what he had found in the Readhimer Mill Branch. He had his pockets full of broken Indian arrowheads and pieces of flint. Envision the Indian making the arrowheads while sitting on the bank above the stream and letting the discards fall in the water.

Both Caroline and Dr. Swanton were thrilled over Richard's find and wanted to be shown where he had found so many. Eight-year-old Richard, very happy to comply, took them to what Jessie says is now called the "low water" crossing of the Readhimer Mill Branch over the Sparta Road.

Jessie says, "Picture this: not only Richard but also the two grown-ups all barefoot in the water patting the sandy bottom with their feet to make the pieces of arrow-making flint rise to the top so they could pick them up." Richard was greatly flattered by all this attention – no wonder he idolized Caroline Dormon!

Richard sold the artifacts he had found to Caroline's cousin, Screvin Sweat, who later turned them over to the Williamson Museum at Northwestern University. Mostly what Richard sold were atlatl points for which he got 25 cents each for perfect ones. Atlatls (throwing sticks) were used by the Indians for hunting large game and also in warfare. To make atlatls the Indians took a stick of about five feet in length and put a point on it made from stone or bone.

Pete Gregory uses the atlatls in teaching classes at Northwestern. The museum there also has some of the baskets that Caroline had which were made by the Chitimacha Indians. The Chitimachas are the only Indian tribe in Louisiana to still occupy a portion of their aboriginal lands, although greatly reduced from what it once was. Now they are found only in an area along Grand Lake and Bayou Teche.

Julie Callihan, a board member of the Foundation for the Preservation of the Caroline Dormon Nature Preserve and the granddaughter of Sudie Lawton who was the foundation's first president, had a reference to Indians in a letter she wrote to another board member. She penned, "When I was a little girl and would go with Sudie to spend the day at Briarwood, they's make me stay on the porch of Miss Carrie's house so I wouldn't get lost in the woods. Miss Carrie told me to stay there and watch for the spirit of the Indians as they wandered here and there. To listen for their songs in the wind and the echo of their footsteps on the pathways. I was always so disappointed to leave and not have seen an Indian, spirit or real life, wander past the porch."

Julie also wrote in this letter, "Bebe (her aunt) and Mama weren't allowed to ride in the car when Miss Gin was driving because she always drove with her legs crossed like she was just sitting down."

HELPERS AT BRIARWOOD

Nora Patterson was a black woman who was Caroline's dependable helper in gardening, one she fully trusted with her plants. In an interview with Rick he said, "Nora was a real sweetheart. I'd often pick her up to take her to town and help her in other ways. Sometimes she'd mistake me for my dad."

Nora once told Caroline, "You know, Miss Carrie, people does say you is peculiar." Fran Holman Johnson included in her book *The Gift of the Wild Things* the following words that Caroline had written which show that she recognized the truth of Nora's statement. In speaking of her neighbors, she wrote, "At last they have just given me up as hopeless. They know I am crazy, but harmless, so anything I do is all right. A really happy state of affairs."

Nora's daughter, Dosia Murphy, worked with Caroline later. As I penned before, Dosia cooked for Caroline as well as working outside. When Dosia was advanced in a pregnancy, Caroline wrote in a letter to Sudie Lawton, "Dosia is going to have that baby in my kitchen!"

As to other work, Dosia said, "Miss Carrie really wants us to pick up sticks outside rather than cleaning house." Richard told me that Caroline didn't want them in her work room as she always had her things laid out just where she wanted them.

At the first native plant sale held at Briarwood in November of 2014, I had the pleasure of meeting Dosia's daughter Eileen and her daughter Faith who now help out at Briarwood. Since Eileen was assisting with the sale of plants and seemed quite knowledgeable about them, I asked her if she had a background in horticulture, to which she replied, "No, but I'm learning." And what better place to learn than at Briarwood.

Jessie told me that Eileen's sister Ruth Ann had worked with her for about 15 years. She added that she remembered Nora as being especially good at working in what is now known as the Bay Garden. The four generations of the Patterson family who have worked at Briarwood are Nora, her daughter Dosia, her granddaughters Ruth Ann and Eileen, and great-granddaughter Faith. Richard, Jessie and Rick speak highly of all of them.

Caroline had this to say about Nora in her diary she began writing in 1937, "Why is it that Nora is the only person that I can trust in my wildflowers? She never pulls up or even steps on one although she may be working in a perfect thicket! Nora is equal to any man – how I envy her that strength! Maybe I could amount to something if I had bull strength."

RICHARD BECOMES CURATOR OF BRIARWOOD

In her very last days, Caroline summoned Richard to her bedside to ask him to take over the care of Briarwood, a position for which she had been molding him for many years. As Richard, Sudie Lawton and Arthur Watson gathered around her bedside, Caroline admonished them, "Don't take time to mourn. Get on with the task at hand."

When Caroline asked Richard to take over the care of Briarwood, she said she would pay him $100 a month. He told me, "I didn't have the heart to tell her that would hardly buy milk for our family." Fortunately, the formation of The Foundation for the Preservation of the Caroline Dormon Nature Preserve, Inc. has given Richard and Jessie the opportunity to see the fruition of Caroline's desire that Briarwood would become a learning laboratory for future generations to see and experience an old-growth forest and its rare plants and animals.

Richard and Jessie have always been strictly volunteers, never accepting any money from the Foundation or even from visitors to Briarwood. The Board members are also volunteers, covering their own expenses for trips and the official meetings. When Richard gives talks to clubs, groups and organizations, he requests that any money offered him should be donated to the Foundation.

Jessie says, "Our goal was to see the Trust Fund with a total of $1,000,000 before we retired as curators and it did reach $700,000 for which we are pleased to have had a hand in accomplishing. The Foundation provided us a place to live and a Ford Courier truck to use; otherwise we paid all our own expenses. As a result of not accepting any money, our rewards have been far greater than money could provide."

Jessie, Richard and Sudie Lawton in front of log cabin

An article in the October 1988 issue of *American Horticulturalist* magazine had this to say about Briarwood after Richard took over its management:

Both he and Jessie devote all their spare time to making Briarwood the kind of nature sanctuary Miss Carrie dreamed of – a preserve not only for plants and animals, a place where the hand of man lies lovingly on the land, letting nature run its course with a minimum of human interference and making only 'improvements' that further Miss Carrie's desire to preserve the southern flora she saw ravaged by private industry during the early part of this century.

Jessie says, "It wasn't until Miss Carrie passed away that we fully realized we had to learn. Earlier much information just went in one ear and out the other." I believe Jessie is selling herself and Richard short in saying "much" information didn't sink in. The place to use the word "much" is in describing the wealth of knowledge Caroline had passed down to Richard and Jessie from which we all now profit.

Caroline had fallen in love with Jessie from the very start, realizing that when the plants were supplied, it was Richard who did the digging but then Jessie took over to nurture them. Richard says, "She has the talent that Caroline had with plants. She can choose just the right plant for the right place. She's very good. But I can dig holes!"

Although Caroline was usually generous to share plants with others, there could be rare exceptions. Once Sudie Lawton and Caroline were walking down the road in Briarwood and Caroline showed Sudie this gorgeous plant, exhibiting much excitement about it. Sudie said, "You have to give me a piece of it," to which Miss Carrie replied, "Sue, are you crazy? I wouldn't give it to God."

In speaking of Caroline, Richard says, "She was the first to suggest and promote the establishment of our state parks. She was a real individualist at a time when people were not prone to do this. She was warm, lovable, extremely loyal to her friends. She didn't put up with anything – you didn't step a foot on Briarwood without her telling you where to step." Then speaking of her frugality, he added, "Her sacrifices were unbelievable. Her austerity sometimes bordered on the ridiculous. She'd spend money on plants before she'd pay her gas bill."

When Caroline wrote of her 'gift of the wild things' in her diary in 1937, she penned, "I'm so sorry for poor things who fret and pine for fine clothes and houses and cars. Smug, am I? No, just humbly thankful."

In an article written by Janie Flournoy for the Shreveport Times in 1978, she referred to Briarwood as "the Noah's ark of plants." Richard says, "This is the place to teach people to like the outdoors." Both of these statements bring to my mind the title of the book *The Last Child in the Woods: Saving our Children from Nature-deficit Disorder* written by Richard Louv. I'm recommending to my pediatrician son that he prescribe a visit to Briarwood for any of his patients he suspects of having this disorder.

Jessie says that had it not been for hungry voles what is now a beautiful meadow of wildflowers at Briarwood might instead be a garden for old heritage roses. In the late 1970s, Cleo Barnwell had formed a group known as the Briarwood Horticulture Society. They were granted permission by the Dormon Foundation to establish a rose garden in an area where Caroline and Virginia had once grown bulbs for sale.

The Horticulture Society purchased about 100 roses and temporarily planted them in another site while Richard did extended work to clear the trees and other growth that had invaded the meadow area. While Richard was laboring, hungry voles were also busy; they ate the grafted root stock of all the roses down to nubs. Jessie says that she and Richard were relieved that instead of a garden of roses, the site became the Wildflower Meadow.

The annual show in the Wildflower Meadow begins in early February with the appearance of the rare luna fern, discovered there by R. Dale Thomas, and the adder's tongue fern. Then the blooming season begins with the bulbs in February and March, followed in succession by masses of Louisiana blue stars that are loved by the butterflies, sensitive briar, skullcap, daisy fleabane, *Rudbeckia alismaefolia,* obedient plant, a few coreopsis, butterfly weed (not nearly enough, Jessie says, to suit her and the monarch butterflies), flat-topped goldenrod (*Euthamia),* liatris, *Baptisias,* goldenrods, and *Silphiums.* Jessie says something is putting on a show in the meadow from late winter into November. In addition to being a magnet for butterflies, the meadow is favored by dragonflies as well.

After the seeds are set in the fall, the meadow is mowed very close. Jessie told me, "The Wildflower Meadow does have its undesirables such as trumpet vine, blackberry vines, tree sprouts and French mulberry, but I recommend planting so much good stuff it will hide the pests. Anyway, the mowing keeps them in check. When I put out a new plant I simply dig up a clump of grass or something undesirable and put my new plant in its place." She added, "The horrid ole oxalis from South America is becoming a nightmare! Would love it if the voles would suddenly find the oxalis to be tasty candy!"

Jessie said, "The Wildflower Meadow has given the deer a perfect place to hide their newborn fawns. More than once we have almost stepped on a fawn hidden away among the plants by its mother." Spotting one of these delightful creatures among the flowers should hasten recovery for any visitors who might be suffering from nature-deficit disorder.

Richard told me that they often have teachers from other states who come to visit Briarwood. One group was asking about the games that were played by children in this area in earlier times. Richard told them about one game in which they would pull a small tree over so that one person could get on it. When it was released it would catapult that person into the air.

A teacher from Illinois then said, "I'll bet another game you played was Civil War," to which Richard quickly replied, "No, we couldn't find anybody to be Yankees." He said the Illinois woman turned beet red, but the other teachers were pleased he had responded as he did. Richard told me, "I didn't mind being a dog, deer, cowboy or Indian, but not a Civil War Yankee."

Full of humor himself, Richard fully appreciates it in others. He says Lewis Grizzard was a favorite after-dinner speaker of his. In one speech Richard heard him make, Lewis said that he lived in Chicago for a year or so before he escaped. He added that they only had two seasons there: winter and the Fourth of July. Other humorists Richard enjoyed were Jerry Glower and Justin Wilson.

Caroline established the wild garden section in Longue Vue Gardens in New Orleans around 1936 using native irises and wildflowers. These gardens, originally owned by Edgar and Edith Stern, are now on the National Historic Register and open to the public.

Richard told me about a visit Edith made to Briarwood in her later years. She was using a wheel chair during the last two years of her life but did not choose to use it this particular day. She wanted Richard to show her around Briarwood but didn't want the others who had accompanied her to be with them. Richard honored her wishes and escorted her by holding her arm to steady her, telling her that he often had carried Caroline piggyback through boggy places. He added that years ago he and a boyhood friend would sometimes form a "chair" with their arms to carry Caroline through such places.

While at Briarwood, Edith wrote a check to cover the pouring of the foundation of the Headquarters Building, later donating more. Richard says, "Even after Miss Carrie died, people have been so generous to go out of their way to help at Briarwood. We couldn't operate without them."

Edith Stern (in car) and Sudie Lawton

One of Richard's first tasks as curator was to clear out the undergrowth that had choked the Bay Garden, preparing it for the many irises that were to come. Jessie and Sudie had sent out requests asking for donations of irises to augment those already in the garden. Due to the response of so many, the Bay Garden is now a spectacular showcase of blooming irises – those native to Louisiana as well as some hybrids – for visitors to enjoy during the bloom period.

Richard and Rick constructing a path across a wet area

In the newsletter for October, 1973, Richard wrote:

> Isn't fall a wonderful season for those who care? The sound of the first bird on the fall migration flight, the aroma of Indian pipes in a pine forest, and the sight of a tree decked in fall colors are a balm to the soul. To walk along a quiet trail at sunrise and feast upon the sight of dew-covered spider webs, and to examine the hundreds of different mushrooms in every color, size and shape on the moist forest floor, to us, this is Life's greatest reward. What a wild beauty, fragrance and melodious harmony Briarwood possesses above all forests!

A photograph of a Ford Courier and a Roto-tiller appeared in the same newsletter with the caption, "Look what your dues have bought." Richard put both the car and the tiller to good use in his work at Briarwood.

Richard working by the pickup

In his newsletter for October 1996, Richard wrote, "Today there is a touch of autumn in the air. I kept looking, sniffing and hoping to see the Indian pipes. There they were poking through the pine straw underneath the tall pines on the hillside. Years ago Caroline wanted some growing near her house so she raked back the pine straw, planted the seeds, replaced the straw and then waited for the results. I'm sure several of you can remember her unrestrained joy when the first clump of Indian pipes bloomed for her."

In her diary written many years earlier on October 1, 1937, Caroline had penned, "Indian Pipes! It's too early, I protested, my feet already following my nose just like a good old bird dog." Caroline concluded her entry by writing, "Oh, strange flower that grows like a fungus from decaying vegetable matter, which has no color but is a true flower and bears seeds. But for all your strangeness you are a lovely and lovable forest sprite. I love the name the little sandhill children give them, 'frost pipes'."

After reading Caroline's diary, Richard concluded his entry about Indian pipes in the 1996 newsletter by saying, "I am amazed at the similarities of today's everyday life at Briarwood. It seems that nature's rhythm never changes. It's unbelievable that almost 60 years have passed since she wrote those words."

RETIREMENT OF SUDIE LAWTON AS PRESIDENT OF THE FOUNDATION

When Sudie Lawton retired in 1997 after 25 years serving as the first president of the Foundation for the Preservation of the Caroline Dormon Nature Preserve, a board member, Marion Bienvenu, wrote the following poem as a tribute to her.

TO SUDIE FROM YOUR BOARD: A LOVING TRIBUTE

A visionary, determined, not taking a "no"

Sudie Lawton has paved the path we will go.

With grit and charm she's helped us conserve

Our matchless heritage – Miss Carrie's Preserve.

Her love of preservation and of beauty

Has made Briarwood's call a joyous duty.

Each time we gaze around the grounds

Sign after sign of her devotion abounds.

To Sudie, our leader, we give our thanks.

With care and perception you filled our ranks.

Heart-filled good wishes from the Board you created;

Your gifts to us are much celebrated.

Marion Bienvenu succeeded Sudie as president of the Dormon Foundation. Her father, the late Arthur Watson, was a charter member of the Board of Directors and served as its treasurer.

FORMATION OF THE LOUISIANA NATIVE PLANT SOCIETY

In 1980, Karlene DeFatta, Dee Bishop and Teresa Thrash had formed a native plant society for the Northwest area which included Shreveport. Richard says he was shocked when he realized that there was no statewide native plant sociey in Louisiana. David Hiekamp, a friend of Karlene's from New Orleans, had wanted such an organization for the state.

In 1983, David and many others with like interests met at Briarwood to form the Louisiana Native Plant Society. Richard says, "Most of the attendees sat on the ground as I didn't have that many chairs."

The newly-formed society chose the central location of Alexandria for their meetings. Around 30 people met at the Extension Office there, where Robert Turley was County Agent, for their first meeting. (I did not attend that first meeting but joined the society later in the year.) Officers elected at this meeting to serve the LNPS were Richard Johnson, president; Ben Martin, vice president; Neil Bertinot, secretary; and Elinor Herd, treasurer. It was decided to have annual business meetings in the winter and to engage in field trips at other times of the year.

For the winter meetings, the LNPS continued to have them at the Extension Office for about 10 years until it became too costly to meet there. After meeting at the 4-H Camp Grant Walker near Pollock for a number of years, it likewise proved to be too expensive. The meetings were then moved to Camp Hardtner, also near Pollock. This camp is the ministry of the Episcopal Diocese of Western Louisiana.

When Richard had asked Jackie Duncan, LNPS treasurer, if the society paid anything at Camp Hardtner, she said it didn't. Richard's reply was, "That's our style." Caroline had been great friends with the Hardtner family for whom the camp was named and Richard's dad was raised about six miles from there.

Richard tells about an incident that happened on a native plant tour in Washington Parish. He said that while on the tour, Robert Murry had a wheel run off of his truck due to the lug nuts not having been put back on tight enough after a flat. Robert took an aluminum wrapper from an ice cream sandwich to use under the lug nuts to secure them on the wheel. Richard told him, "Yeah, yeah, you're going to end up in the ditch," but he made it. Richard told me, "I learned something new. Being the son of a hill country farmer, I thought I knew just about everything."

In a winter meeting of the LNPS in 2015 at Camp Hardtner, Richard gave a program entitled "History of Briarwood and Associated History of LNPS." In his presentation he paid tribute to several past and present members. Of the late Carl Amason, he said, "Carl was one of the greatest plantsmen I've ever met. He loved to come down from Arkansas to our native plant meetings."

Then he spoke of John Mayronne who has a landscaping and native plant nursery at Covington. Richard said "John has made certain that we had all these native plants; he knows their history. He has been a great motivator for us."

Richard had this to say about Annette Parker, "I think of Annette as having been a great teacher and also as a plant person." He said that Annette had to put fences around certain plants because of a major deer problem, but she had the perfect solution for other problems. She says, "If it won't grow, let somebody else tackle it."

In the brochure prepared for the LNPS 2015 winter meeting, one of the things the current president, Dr. Gladden "Bud" Willis, wrote about Richard was, "After Caroline's death and with the establishment of the Foundation for the Preservation of the Caroline Dormon Nature Preserve Inc., Richard began a labor of love and dedication to the preservation and growth of Briarwood that continues to this day." Bud is a retired surgeon who currently operates Willis Farms, a native plant nursery near Doyline, Louisiana. He is also the current president of the Dormon Foundation.

I founded the Haynesville Celebration of Butterflies in 1999 and continue as the director of what became an annual event. Many of the LNPS members have been speakers at the festival through the years, among them Richard, Carl Amason, Bill Fontenot, Charles Allen and Jeff McMillian who operates Almost Eden, a native plant nursery near Merryville. Most of these have graciously returned a number of times. Charles Allen appears every year as a speaker, giving programs on various subjects. Richard, assisted by Jessie, has also given numerous presentations.

Richard says the spraying of highway roadsides got him "real unhappy" because it destroyed the wildflowers growing there. He said, "Whether the plants lived or died was not considered by the workers to be their problem. All they wanted to do was put in their eight hours and go home to their wives and children. And appealing to the ones in charge didn't help either."

Back before the days of Lady Bird Johnson's success in having native wildflowers established along the highways of Texas, Louisiana was the wildflower state of the nation. At that time a scythe was used for cutting the plants along highways after the flowers had set seeds. This permitted the wildflowers to come back year after year.

Two charter members of the LNPS, Ben Martin and Beulah Bergeron, were especially good friends even though Beulah was much older than Ben. He'd say such things to her as, "Come on, Ole Woman (his pet name for her), let's go look at flowers."

Beulah was a long-time garden club member. Once when Richard was giving a program to an Alexandria garden club, following his presentation one of the members there led him to the nursing home where Beulah was a resident. Richard says, "Beulah let me know I was running late, but she forgave me when I gave her a bottle of wine. I always managed to bring a little 'refreshment' for her even though it wasn't allowed." He said he would smuggle the wine in with flowers and other such gifts.

WILDLIFE AT BRIARWOOD

Caroline often wrote in her diary of the wildlife at Briarwood. In June of 1937 she made this entry about a titmouse she spotted on her porch, "There was a comical little titmouse with a mouthful of cedar bark. This he was holding down with his foot and shredding it with his beak. He looked as if he had a fierce mustache. And he wasn't even satisfied yet, but hopped up on the post and began pulling off more bark. I could not follow him with my eye to see where his nest was under construction; in some hollow limb close by, I know. I have often caught titmice pulling the hair off the hide-bottom chairs on the porch!"

In her diary in May of 1938, Caroline again mentioned the titmouse collecting nesting material from the hide-bottomed chairs and the bark from the posts on the porch. But she wrote of another source for such material, penning, "Today I was sitting very quietly on the log seat between two of these posts, looking at a book, when a titmouse plumped down on the floor in front of me. He tilted his eye up at me and said, 'chi-cha-cha-cha' in a most impudent manner. Then he flew from post to post, making a survey from every vantage point. Evidently my uncombed and fuzzy hair looked useful, for he planted himself firmly on my head, braced his feet and began pulling it out by the beakful! 'Pop' went the roots as they came out of their tiny sockets. I suppose I would be wearing a wig but I just had to call Sister to come see (I wanted a witness to this performance), and the robber flew away. I wish I could see that nest, but as they build in a hollow, that would be difficult."

Caroline was always interested in the many birds at Briarwood. The writer, Lyle Saxon, addressed this in his poem about Caroline, entitled "Girl in a Tree." Two lines of the poem are:

> There she sits and juggles words
>
> And nibbles twigs and watches birds.

Richard is very protective of the plants and animals at Briarwood. Once when his grandson, Luke, was a nine-year-old, Richard told him to return a lizard to the exact location where he found it. He explained to Luke, "Every lizard has his own territory. If you take him out of there, why, he'd be lost. It'd be like if we picked you up and dropped you over in Georgia."

In the newsletters of Briarwood Richard often wrote of the varied wildlife to be found in the preserve. In the October issue of 2001, Richard wrote:

> Just yesterday I noticed a box turtle (terrapin to us) happily munching on a white mushroom. Conditions have been just right to bring up all kinds. We have enjoyed looking at them. The squirrels like to upend the mushrooms on stumps, the picnic tables, or even in the middle of the paths, leaving them to cure for about three days before eating them. I have often marveled at how they developed this method of food processing.

A story of great interest is this one Richard wrote for a newsletter:

> Over the years I have shared many stories about our wildlife but this is one of the best. This old male raccoon that we call Lop-ear is a veteran of many fights. He had become a problem in the garage, taking the lid off the food barrel and helping himself. One night in April I was working in the shop when I felt a tug on my pants leg. It was Lop-ear wanting me to stop and feed him. So into the live trap he went and I hauled him 20 miles east across two creeks and let him go. My problems were over, I thought! Three weeks later he came bounding down the path, reared up on the correct glass door and demanded food. He sure knows where to find a 'happy meal.' He is doing great but has left us wondering how did he do it?

In the newsletter of July 2011 Richard penned, "It looks like a summer of three digit temperatures has arrived! As if a very dry and the hottest June ever for us was not enough to struggle through. I remember the Dust Bowl days of 1935 when we ate dust blown in from the high plains of Texas and Oklahoma with every meal. But we received more rain then than we have in the past 12 months."

During a dust storm Richard's maternal grandfather exclaimed, "Here comes Texas!" Richard says, "It was West Texas, it just came to us." He remembers seeing a huge black cloud that looked like it was rolling with the wind. (Since I lived in West Texas during the Dust Bowl days, I can readily assure you that we did not send all of our dust to Louisiana!)

Concerning drought at Briarwood, Caroline once wrote in a letter to Sudie Lawton, "It rains circles all around me – I get only sprinkles. The Lord must have turned me over to the devil 'for a season' as he did poor Job!" Yet even in the midst of the terrible heat and drought of 2011, Richard found enjoyable things of nature to write about in the July issue of the newsletter:

> One of summer's great joys is walking in the forest. Pick a path underneath the trees from daybreak to sunrise and meander along taking in all the sights and sounds. Isn't it amazing how the forest comes alive! For almost an hour bird calls welcome the new day and the squirrels come to see what all the fuss is about. This year we have two wood thrushes starting the morning chorus keeping it up for most of the day. The other best time is late evening after the sun has sunk low in the west and we are treated to the day's last hurrah. Then bird calls seem to be discussing their day and as darkness approaches the tree frogs challenge the chorus frogs to see which can sing the loudest. Not to be ignored, the katydids by the hundreds add their sound to the melody. Lightning bugs add an enchantment to the evening with their flashing lights that rival the stars above. This draws us outside each summer for there is no greater sight or sound to be witnessed!

One may think of Briarwood principally as a place to visit to view the plants, but it is also a true sanctuary for birds. Richard often writes about them in his newsletter such as appeared in the April issue of 2013:

> Today we watched our feathered friends getting ready for that long flight to their nesting grounds. I'm sure our resident birds will be glad to see them go. We saw more pine siskins this year than in several winters but now our white-throated sparrow count is low. Just the other day we could hear them singing and looking to where they feed on the ground could see a group of eight males busily scratching

for feed, clad in their spring finery. I have missed the passing through of the whip-poor-will for the past two years. They are intent on reaching their nesting territory so do not tarry long. Our resident chuck-will's-widow has become rare – we think due to fewer farms in the area and more trees which do away with the open spaces for catching insects while in flight. On our way to Ruston I saw a bald eagle fly up from roadkill. This was my first-ever sighting. I understand that an eagle nest has been found in our area.

Caroline's great interest in birds was evidenced by her book *Bird Talk* published in 1969. In it she has a chapter on the IQ of birds in which she discusses a bird's facial expression as being characteristic of the species. She wrote, "Compare the bright inquisitive looks of chickadees and wrens with the placid gaze of the thrushes. Thus birds display their intelligence or lack of it. The dull stare of a robin suggests stupidity. (Of course I would be tarred-and-feathered for this in the North!)"

Some of us in the South might join in the tar-and-feathering did we not have such a great respect for Caroline. I think most of us enjoy robins and find it hard to understand Caroline's loathing of them. Once when Caroline sighted a flock of robins eating berries on her shrubs, she shouted, "You Yankee robins, go back where you belong!"

Richard continues to share with us news of the wildlife at Briarwood. In the July issue of the newsletter in 2013 he wrote:

> Even summer nights have much to offer. The chuck-will's-widow is still calling from our neighbor's cow pasture to the north and the barred owls can be heard calling their young to come catch a frog (or hopefully a pesky vole). At dusk we are treated to the ritual flight of the lightning bugs (fireflies to you) a sight everyone should have a chance to enjoy.
>
> I wish we could have shared with you the privilege to witness the raising of a young red-shouldered hawk. This one stays near our lawn where it finds lizards and toads to munch on. It changes perches occasionally but prefers a six-foot high post. The parents go out in in the woods to find

mice, rats and other small animals to bring back to it to eat. I can tell you they are ready for it to be independent!

Numerous grapes of various species grow at Briarwood and in the surrounding woods, although clearcutting in the late 1970s destroyed many of the vines outside of Briarwood. Jessie says that sandhill grapes, which are much like the Concord, make great juice. She said her mother made "medicinal wine" from grapes and that she was not the only one at that time to do so. Jessie spoke of taking large wash tubs to the woods to put the gathered grapes in. Summer grapes were one of the species harvested early in the season with muscadines ripening later. Richard said that the muscadines were his favorite and that he would throw up pine knots to knock the grapes to the ground.

Richard wrote the last of his newsletters in October of 2013 just after Rick had taken over as curator in August of that year:

> A few weeks ago our summer helper, Chris Blanchard, saw six little turkey poults near Wings Rest pond. I heard both the gobbler and hen call back in late May on the east side of Briarwood. The Wildflower Meadow is a perfect nursery for them with lots of insects and provides dense cover from the air and ground. Rick Johnson, our new curator, saw his first prothonotary warbler when it visited the water feature he installed back of the Headquarters Building. Back in the early 1970s we had one build in a gardening can hung in the doorway of a seldom-used shed. A surprise because we thought the prothonary only nested in cavities near water. Its brilliant yellow feathers sure get your attention in a hurry.

Like father, like son, Rick continues in his dad's footsteps in writing about the wildlife at Briarwood. Unlike Caroline, Rick doesn't seem to mind the robins eating Briarwoods' berries. Read what he has to say in his newsletter for January of 2014:

> On October 20 we knew that fall was fast approaching as we heard, then saw, the first white-throated sparrows feeding back of the headquarters building. Then as if to add an explanation point to the statement that winter was on its way, on November 7 while walking along the entry road I observed the first robins of the season feasting on the berries

of the dogwoods. That same night I stepped outside to look at the beautiful star-filled sky and was greeted by the sound of a flock of geese talking as they winged their way south . . . a wonderful sound I remember well from my childhood growing up in this area.

Now that we've had a period of cold weather we are enjoying our northern visitors at the feeders – the juncos, chipping sparrows, nuthatch and goldfinch have joined the resident cardinals, chickadees, titmice and blue jays to provide a colorful scene.

As American robins are permanent residents in the south, I asked Jessie why Rick would refer to seeing the first robins for the season as I thought they would have been there all year. She says robins do not nest at Briarwood, seeming to prefer the habitat of the nearby towns during nesting season. Richard added that he recalled having seen robins in the fall among the large flocks of blackbirds feeding in the fields.

At the spring picnic in 2014, Rick conducted tours of Briarwood. He had this to say, "A special moment took place for me during the morning walking tour when a male ruby-throated hummingbird began to feed on the red buckeye just as I began to explain to the visitors how their northern migration is timed to the blooming of the buckeye. His timing was perfect!"

Because of such an interest in the birds of Briarwood, at the annual spring picnic there in April of 2015 a birdwatching tour was led by an avid bird expert of the area, John Dillon. There was a barred owl that seemed to be taking the tour with the group, sighted and heard in many areas along the route through the woods. Rick had written about barred owls in the newsletter of the previous April:

On January 28 I went out to look around the preserve after our second snowfall and caught sight of one of our barred owls in flight. From the frequent calling, I believe that they have nested somewhere in the Cow Oak Flats area. It is really special to hear them hooting and caterwauling; sounds like they are discussing the upcoming night's hunt and laughing about it.

In the newsletter for July, 2015 Rick wrote of the magic of evenings at Briarwood:

> Southern summer evenings are a truly enchanting time. As the sun releases its grip on the day and slides below the western horizon the magic show begins. First, the scent of Magnolias begins to perfume the air with a fragrance that is far superior to anything that man can make. This followed by the night's orchestral entertainment comprised of chirping crickets, buzzing cicadas and katydids with bass accompaniment provided by bull frogs. In that last bit of twilight, just before full night, the light show begins courtesy the lightning bugs as they ascend from the forest floor blinking their signals to one another. When night takes over from the day and the rising moon casts its light over the land, more creatures add their voices to the melodious sounds of the night. Leopard, toad, tree, and peeper frogs raise a cacophony of sounds while the barred owls hoot and caterwaul to each other as they search for something to feed their ever-hungry young. Occasionally there is a special accompaniment from coyotes, starting out as a single soulful howl but soon joined by the yipping and howling of others as they join up for their night search for food.

Richard says the first armadillo for the area was sighted by his Uncle Steve Johnson in the 1940s. After dogs had run the armadillo under the house, Steve retrieved it and put it in a barrel. Richard told me everybody came by to view it as they'd never seen one before. The local name for the armadillos became "grave diggers" because of their habit of burrowing in search of grubs and roots for food.

I've read reports that the first armadillos came across the Sabine River from Texas to Louisiana about 1936. As these animals cannot swim, they walk across on the bottom of a stream or another way to get over it is to gulp large amounts of air to inflate their lungs and intestines in order to be able to float across a body of water.

STUDENT HELPERS AND OTHER WORKERS

Through the years, students from different high schools around the area have helped out at Briarwood. In the summer of 2011, seniors Marlanna Cimino, Zack Friday and Kenneth Sepulvado, all from Lakeview High School which is near Campti, helped Jessie in such areas as the Bay Garden. Parker Johnson, a freshman from Saline High School, assisted Richard in road and trail building plus removing excess American hollies. All pitched in to get water to the plants in what was proving to be a very hot and dry summer.

Jessie in the Bay Garden

In the summer of 2012, student helpers at Briarwood were seniors Nick Cimino and Chris Blanchard from Lakeview High School. Richard said he taught them "Trail Building 101." As Richard brought in different types of dirt and gravel with the front-end loader of the tractor, the students shoveled, raked and packed it in place leaving a smooth surface for the trail. That summer they finished the Cow Oak Flats and the Mill Branch trails and did much work on the Sparta road.

In anticipation of his and Jessie's retirement the next year, Richard wrote the following in his newsletter for July of 2012:

We have gained another wonderful worker-friend, a native New Jersey retired marine, Feliz Rivera. Felix took to a chain saw like a duck to water and no type of work has yet to faze him. He and Rick make a great pair when it comes to clearing undesirable brush. Rick and Denise, whom most of you have met, will take our place here next July, 2013 and Jessie and I will drop back to a supervisory position. This will be the computer generation where I am stuck with my rolodex! As we transfer over 40 years of hands-on information we'll be here for work or consultation to keep it going as Caroline Dormon would want it to be.

RETIREMENT PARTY FOR RICHARD AND JESSIE

A retirement party was held at Briarwood for Richard and Jessie on August 3, 2013. James Durham, representing the Dormon Foundation, gave the following speech:

> The Foundation for the Preservation of the Caroline Dormon Nature Preserve was incorporated September 21, 1971 by Sudie Lawton, Mrs. Jo B. Ducournau and Mr. Arthur C. Wilson. The purpose was to preserve woodlands and the forest lands in their natural state, particularly the 120 acres owned by Caroline Dormon and known as "Briarwood." Richard and Jessie Johnson became the curators in 1972. Most of what the Dormons had developed at Briarwood had deteriorated completely. Richard and Jessie have carried out the improvements that we see today. The foundation has added several pieces of property to the original 120 acres donated by Ms. Dormon and now has about 180 acres.
>
> Mr. Watson, not long before his passing, was working on setting up a permanent endowment fund for Briarwood and wrote letters requesting help from several people. In a letter dated June 18, 1984 to Mr. Harry Balcom of Shreveport, Mr. Watson said, "At the present time we have Richard Johnson and his wife Jessie who are up next to heaven, I guess. We pay them nothing at all except to allow them use of a part of the headquarters building for their home. They even pay the electric bill on the house. They are dedicated people however and I must admire them all the way. However, should Richard and Jessie pass on or decide to leave for any reason, we would really be in a pickle. For that reason, we are working hard to get $250,000 as a permanent endowment. With just the interest on this we would be able to hire a man and woman both to look after the place. We would never get anyone with the dedication that our present curators have but it would at least keep the preserve going."
>
> The endowment trust fund now has over $700,000 and I too must admire and thank Richard and Jessie all the way. They

have dedicated their lives for the last 41 years to Briarwood and created this fabulous place. Now we are fortunate to allow Richard and Jessie to become "curators emeritus" and transition to their son Rick and his wife Denise as caretakers of Briarwood.

As another part of the retirement party ceremonies, James Durham gave the following speech:

> We would like to present these two cypress benches – one for Richard and one for Jessie to be placed side by side at the location of their choice, as they have worked side by side all these years to make Briarwood what it is today. They are made from cypress to hopefully last a long time as a reminder to everyone who visits Briarwood of the lasting dedication Richard and Jessie have made to Briarwood. (The benches have been placed in the pavilion.)

Richard & Jessie seated on donated bench

Both of these individuals have dedicated their lives and worked together these many years to create with their own time and energy and resources what we have today at

Briarwood. When they started there were just a few outside resources from the "Friends" group to help them. Over the years they have used ingenuity and hard work and have nurtured relationships and worked with people, who, in turn were inspired to try and help them. I have kept the books for the foundation for a long time and it has been my privilege to visit them regularly. They have never asked for anything for themselves and only asked for funds to be spent on Briarwood when they couldn't do something by themselves.

Richard wrote this about the retirement party in his newsletter of October, 2013:

> August 3, 2013 we will remember always. What a wonderful retirement party the Board of Directors gave us on that day. The monumental planning that went into making it a success is mind boggling, not even a minute detail was forgotten. We could feel the love flowing from the gift of two benches to Marjorie's floral arrangements that echoed the flattop goldenrods in bloom in the Wildflower Meadow! Jessie and I were so happy and honored to see so many friends in attendance on such a hot day but the best of all was the chance to visit and talk to everyone. You have written cards and letters through the years saying how much you appreciate the work we have done in preserving Caroline Dormon's Briarwood but for you to say so in person and by speeches given in our honor made it very special indeed! Thank you!
>
> It seems just yesterday – not 41 years ago – that Jessie and I assumed the responsibility of caring for Briarwood, Caroline's legacy to nature lovers worldwide. Like Dr. J.K. Small, we were amazed at the variety and number of rare plants Miss Dormon had introduced to her nature preserve, the most notable being the rare *Torreya taxifolia,* which she grew from cuttings from Florida. The following is a quote from Caroline: "The most exciting find I ever made was a *Magnolia pyramidata,* made while on a camping trip in Sabine Parish with my young nephews." There are seedlings growing at Briarwood from those trees, one of which was just found by Rick growing near the new pavilion and is now

20 feet tall. So the discoveries continue; each day brings something different. We owe so much to Caroline Dormon and her vision of preserving this place for the future generations to learn the meaning of ecology and the importance of maintaining diversification in our world.

Richard concluded his message in the newsletter by extending a welcome to visit them at Walnut Hill where they now live. To reach their new home, veer to the right at the front gate to Briarwood, go up the hill and the house is on the right. Richard says, "We're not the curators but we still volunteer to work. I know Caroline Dormon is beaming approval at the change." And having Rick and Denise as the new curators certainly seems the ideal choice to meet the approval of all of us with an interest in Briarwood.

In the first newsletter written by Rick as curator, he penned:

> August 3, 2013 is a day that I will remember always, first for the outpouring of love for my parents as they "retired" from Briarwood and secondly as the day that I took over the reins and began the process of learning the daily operations of Briarwood. To those of you who know my parents well, you know that they really haven't truly retired, just shifted gears. It has been wonderful having them nearby to provide experienced advice.

SCHEDULED EVENTS AT BRIARWOOD

There are several events that occur regularly at Briarwood. Tom Sawyer Day is held twice a year the last Saturday of the months of October and February. Richard says, "We can always use volunteer help, so if you have a group or would just like to come by yourself, we'll gladly hand you a pair of work gloves and let you invest a little 'sweat equity' into the soil of Briarwood."

Those coming for Tom Sawyer Day work in the morning then tour the preserve in the afternoon. Workers are encouraged to bring a sack lunch which is usually augmented with soup or gumbo prepared by Jessie. Yum, yum! Tom Sawyer himself would have relished a bowl of Jessie's cuisine after whitewashing that fence. Aunt Polly was probably too exasperated with him to cook for him that day.

There is a fund-raiser picnic held at Briarwood on a Saturday in early April each year. A delicious meal is served and entertainment is provided by the Back Porch Band. Those attending can go on guided tours or wander off on their own to enjoy the springtime beauty of the preserve.

Another event for Briarwood was added in 2014 with the first annual native plant sale being held on November 1 of that year and the second on November 7, 2015. Many plants grown at Briarwood are available for sale as well as some from Willis Farms and sometimes other native plant nurseries.

At the spring picnic in 2015, announcement was made about a chest being donated to Briarwood by a member of the Back Porch Band. There is a rather lengthy story concerning this chest. A huge old pine tree had fallen across the Sparta Road after a heavy rainstorm had uprooted it. Richard cut the pine up in short lengths so he could move it to the side of the trail.

Michael Yankowski, an art teacher at Northwestern State University, saw the wood and marveled at its beauty. He counted the rings, determining that the tree was 130 years old when it fell. Since Michael worked with wood, Richard and Jessie offered him the pine for his crafting, glad that it would be used for art. It was from some of this pine wood that Michael made the chest. He offered to give the chest to Briarwood if $500, designated as for the chest, was donated to the preserve – he did not want any money for himself. Funds were donated and the chest is now on display in the log house.

Chest made by Michael Yankowski

In addition to the annual events, Briarwood is open for visitors at other times. Richard says, "Come anytime March through June and the months of October and November. Just let us know you're coming or otherwise we might be out chasing wildflowers."

ACCOMPLISHMENTS AND PLANS FOR THE FUTURE

In a letter from Bud Willis early in 2015, he tells of plans to have a boardwalk constructed to connect the Charlotte Collins Trail with Cow Oak Flats Trail. The boardwalk will be a memorial for George K. Gilbert who visited Briarwood many times with his wife Edna. When George passed away, Edna requested that all memorials be sent to the Caroline Dormon Nature Preserve. Richard and Jessie decided that these funds would be used to construct the boardwalk.

Charlotte Collins first visited Briarwood in 1979, finding it such a noteworthy preserve that she willed money to the Foundation. These funds were used to construct the Charlotte Collins Trail (mentioned in the above paragraph) and have it named for her.

In the letter written by Bud Willis, he told of much progress having been made in the previous year. Among work done was the restoration of the Readhimer Mill Pond which was first constructed circa 1850. This pond was named for the original owner, Jack Readhimer, who had a grist mill on the site. Jessie says that they still have parts of the mill in the area near the pond. After Mr. Readhimer passed away, the pond was used by people in the community as a place to swim. It then became known as the Readhimer Community Pond.

Shortly after they married, Richard's parents rented the little house that the grist mill overseer had lived in. Later they moved onto property where Richard's maternal grandparents lived; eventually buying this 80-acre farm and raising their family there.

In 1929 some boys cut the dam on the Readhimer Pond hoping to capture the fish but the water gushed out so fast that all the fish were washed downstream. The result was no fish for the vandals and no pond left for community swimmers. Years later, Richard and two other teenagers spent all of one Saturday digging dirt to fill in holes on the dam. After much excavation that day a heavy rainstorm occurred in the night. Richard says, "All our work went downstream."

On a tour led by Rick during the spring picnic of 2015, he took us by the restored Readhimer Mill Pond, pointing out the large pit by it that had been there many years. The pit remains as a testimony to what proved to be fruitless labor by three industrious teenagers.

Another pond that is on the same stream of water as the Readhimer Mill Pond was also restored in 2014. Originally constructed around 1950 by Richard's uncle, Steve Johnson, it is called the Johnson Pond and is now owned by his heirs. Although it is on adjoining property, water from it backs up into Briarwood. The Dormon Foundation paid for the restoration work on the pond as it is an excellent source of irrigation water for Briarwood plants. Richard says that the Johnson Pond has beavers in it, then adds, "But I don't think we'll have any alligators."

Knowing that water would flood some Briarwood property, Steve consulted Caroline and Virginia about his plan to build the original pond to raise shiners to sell as fish bait. The Dormons gave their approval and Caroline proceeded to move her plants such as mountain laurel and Louisiana irises from this site – it had been called the Beech Garden – to the newly-constructed Bay Garden.

NOTES FROM MEMBERS OF THE DORMON FOUNDATION AND LNPS

I have received a number of notes from members of the Foundation Board and the Louisiana Native Plant Society in which the writers express their respect, admiration and love for Richard and Jessie. James Durham had emailed me copies of the two speeches (printed earlier in this book) he made at the retirement party for Richard and Jessie. In this email he first thanked me for writing Richard's biography and then had this to say about Richard and Jessie, "Two more dedicated people do not exist."

Another board member, Julie Callihan, sent an email in which she wrote, "I have been going to visit Miss Carrie and Briarwood since I was born. To me, it seems as though Richard and Jessie have always been a part of Briarwood. I do remember for certain standing on Front Street (in Natchitoches) with Sudie Lawton and Evelyne Taylor and Sudie saying she had found the most wonderful young couple to come take care of Briarwood and she hoped they would stay forever."

Chris Evans, also a board member, sent me the following information:

> I don't have any particular stories about Richard except to say that over the years I taught, I brought a number of groups out to Briarwood to learn about plants. We'd pair the girls with Jessie and the boys with Richard. He always showed them the most interesting stuff (the wet dog tree was a favorite) and kept them entertained with the plants and also his stories about working with Miss Caroline as a boy. He always knew just what they would like to hear, the only exception being the old tools on the porch at Miss Caroline's house. He'd go on and on about them but by this late in the tour the kids were sort of brain dead, so they rarely got out of his information what he was putting in. Jessie and Richard were always so welcoming and warm to the school groups. I hope in your biography you'll get all his "Miss Caroline stories." He's got really great ones.

The following is something that has appeared on the Briarwood brochure, written by fourth grader Amelia Hall, a student of Chris's at NSU Elementary Lab School who was among those who visited Briarwood:

> Thank you for teaching us about Briarwood. I enjoyed the trees that smelled good all over! I've seen big trees, but until today I've never seen one as big as Grandpappy! The Jack-in-the-pulpit plants were neat and so were the plants that look like dresses. My favorite plants were the mayapples, the plants that eat insects, and the yellow plants that grow in water. They remind me of lanterns. Thank you again.

An LNPS member, Linda Adrion from Shreveport, sent me the following information:

> My mother, Marie Landers, and I attended and enjoyed many a Tom Sawyer Day. We were invited to spend the night with the Johnsons, a great treat for the both of us. Back then Richard fed the raccoons on the back porch of the headquarters building. Since Mom always insisted we "do our part," we brought coon food. Richard served Old Roy dog food, vanilla cream cookies and Moon Pies, all readily available at Walmarts.
>
> We looked forward to seeing the old mama coon eat the Moon Pies. She was the only one Richard fed this truly southern cuisine, the others ate dog food and cookies. There came a day when he fed them less and further away from the house until he stopped feeding altogether. When I asked him why he stopped, he replied, "I didn't want to make welfare raccoons." As many times as I've listened to him, I've never heard the same tale twice.

It may be that Linda just didn't recognize having heard the stories before. Perhaps Richard takes the advice of Mark Twain, "Never let the facts stand in the way of a good story." Actually, Richard has such a talent for storytelling that he doesn't need to alter the facts to make a good story. And if he repeated one, it's likely he'd have thought of some additional things to enhance the telling so you'd be hanging on every word.

Charles Allen, another LNPS member, had this to say about Richard, "Twas sitting with him and someone was complaining about something and Richard told him that if he wanted sympathy he should look it up in the dictionary." Charles added that Richard is known for coining his own common names for plants although he does know both the scientific and recognized common names.

Another thing Charles mentioned was that Richard told him about trying to preserve winged sumac fruits to make tea later, but that it didn't work too well. Charles, the co-author of *Edible Plants of the Gulf South,* is an expert on such things. He is well-known for giving presentations on edible plants for which he prepares a large selection for the audience to taste, including many teas made from plants.

LNPS member Gail Barton interviewed Richard in December of 1995 for a gardening newsletter. Following are some of the things Richard told her:

> We used the road that goes through Briarwood to go to the community store. I would always listen for Miss Carrie. She would be whistling, imitating birds, or talking to her helper Norvella or to the dogs and cats. I would follow the sound to find her and she would always have something to show me or to tell me.
>
> In the 1930s during the Great Depression, Briarwood's wild gardens were a 'native plant nursery.' Miss Carrie liked me because I could climb. She told me that I was the only one who could climb a tree better than she could. In the winter, greenery was collected at Briarwood to be taken to Shreveport where Caroline's sister-in-law Ruth Dormon had a greenhouse. I was the one who could climb up and get the best branches. They sold holly, mistletoe and other greenery but the prime things they sought were *Smilax lanceolata* and cedar with berries. I was so thin and light that I had no problem climbing to the highest branches.
>
> As I grew older, Miss Carrie discovered that I could tell the difference between cardinal flower seedlings and weeds. She paid me to collect seedlings in a shoe box lid and bring them to her. She sold them in Shreveport for 20 cents each even during the depression. As I recall, she paid me 5 cents

for each plant I collected. There were times when I brought home more money than did my father from his work. The Dormon sisters were considered by some to be 'curious' as they did not raise cotton and they sold weeds!

CONCLUSION

Rachel Carson who authored *Silent Spring* in 1962, shared Caroline's views on the wonders of nature. She penned, "A child's world is fresh and new and beautiful, full of wonder and excitement. It is our misfortune that for most of us that clear-eyed vision, that sure instinct for what is beautiful and awe-inspiring, is dimmed and even lost before we reach adulthood. If I had influence with the good fairy who is supposed to preside over the christening of all children I should ask that her gift to each child in the world be a sense of wonder so indestructible that it would last through life, as an unfailing antidote against the boredom and disenchantment of later years." Treatment for nature-deficit disorder – as mentioned earlier in this book – should assure the fulfillment of Rachel's wishes.

In the newsletter for January of 2000, Richard had written:

> Envision Caroline Dormon as she walked these woods one hundred years ago exclaiming over some treasured plant on the forest floor, singing with the birds in the tree tops or taking a second to hug a tree that today's children view with awe. The huge oaks along the fence rows and Grandpappy, the 300-year-old longleaf pine, have seen many generations pass. We know that with our help and yours they will see many more.

Now envision Caroline beaming down on her protégé, Richard, pleased with all that he and Jessie have done to augment and preserve the beauties of her beloved Briarwood. Listen to the sounds of the chirping birds, the chorus of tree frogs, howling of coyotes, trickling of refreshing streams, and the cheerful voices of visitors to this bit of paradise on earth. View the beauty of the plants there that offer their blooms and changing foliage in timely procession through the seasons as God has planned for nature's variety show.

Rick refers to Briarwood as "an ever-changing, but in the essentials, a never-changing place." With him as the present curator following the dedicated work of Richard and Jessie for those 41 years they served in this position, we are assured that Briarwood will remain the "bit of heaven on earth" that it has been for so many decades.

Briarwood is the place to visit to enjoy both the southern hospitality of Richard, Jessie, Rick and Denise as well as the beauty of the diverse collection of native plants there. And remember that Richard and Jessie are now living right outside of the preserve, helping to assure that Briarwood will continue as a special place that attracts visitors from around the world.

Richard, Jessie and Rick seated on cypress swing donated by Loice Kendrick-Lacy

www.ingramcontent.com/pod-product-compliance
Lightning Source LLC
Chambersburg PA
CBHW041604110426
42742CB00043B/3451